21st International Conference on Laser Spectroscopy

(ICOLS 2013)

Journal of Physics: Conference Series Volume 467

Berkeley, California, USA
9-14 June 2013

ISBN: 978-1-62993-446-4
ISSN: 1742-6588

Printed from e-media with permission by:

Curran Associates, Inc.
57 Morehouse Lane
Red Hook, NY 12571

CURRAN ASSOCIATES INC.
proceedings
.com

Some format issues inherent in the e-media version may also appear in this print version.

Printed by Curran Associates, Inc. (2014)

Institute of Physics
Dirac House, Temple Back
Bristol BS1 6BE UK

Phone: 44 1 17 929 7481
Fax: 44 1 17 920 0979

techtracking@iop.org

Additional copies of this publication are available from:

Curran Associates, Inc.
57 Morehouse Lane
Red Hook, NY 12571 USA
Phone: 845-758-0400
Fax: 845-758-2634
Email: curran@proceedings.com
Web: www.proceedings.com

TABLE OF CONTENTS

Laser Spectroscopy and Frequency Combs

Theodor W Hänsch and Nathalie Picqué
Max-Planck-Institute of Quantum Optics, Hans-Kopfermann-str. 1, 85748 Garching, Germany
Ludwig-Maximilian University, Faculty of Physics, Schellingstr. 4, 80799 Munich, Germany

E-mail: t.w.haensch@mpq.mpg.de

Abstract. The spectrum of a frequency comb, commonly generated by a mode-locked femtosecond laser consists of several hundred thousand precisely evenly spaced spectral lines. Such laser frequency combs have revolutionized the art measuring the frequency of light, and they provide the long-missing clockwork for optical atomic clocks. The invention of the frequency comb technique has been motivated by precision laser spectroscopy of the simple hydrogen atom. The availability of commercial instruments is facilitating the evolution of new applications far beyond the original purpose. Laser combs are becoming powerful instruments for broadband molecular spectroscopy by dramatically improving the resolution and recording speed of Fourier spectrometers and by creating new opportunities for highly multiplexed nonlinear spectroscopy, such as two-photon spectroscopy or coherent Raman spectroscopy. Other emerging applications of frequency combs range from fundamental research in astronomy, chemistry, or attosecond science to telecommunications and satellite navigation.

There could be no laser spectroscopy without the laser. In 1964, the Physics Nobel Prize recognized the seminal work on masers and lasers by Townes, Basov, and Prokhorov. Lasers soon became intriguing tools for nonlinear spectroscopy. Laser spectroscopy really took off with the advent of widely tunable and highly monochromatic dye lasers [1] in the early seventies. The first ICOLS conference was held in Vail, Colorado, in 1973, organized by Richard. G. Brewer and Aram Mooradian. Since then, the field has attracted many talented scientists, and today we count no fewer than 17 additional Nobel Prize winners in our community.

Sometimes, relatively simple concepts can create new tools that open new doors. One example is the laser frequency comb [2], which arrived about 15 years ago, as a tool for measuring optical frequencies of 100's or even 1000's of THz. By providing a phase-coherent link between the optical and the radio-frequency regions, a frequency comb makes it easy to count the frequency of a laser wave. Before, this task had remained close to impossible. A frequency comb is produced by a mode-locked femtosecond laser. Such lasers have been developed since the late 1960s, and they were in widespread use as sources of short flashes of light for the study of ultrafast phenomena in molecules, liquids, and solids. Researchers interested in precision spectroscopy were using a different kind of laser, operating on a single resonator mode, and perfected to achieve very high stability. Since then, frequency comb techniques have greatly enriched and extended the frontiers of both fields. A mode locked laser is a source of a million perfectly evenly spaced comb lines or longitudinal laser modes, all

Journal of Physics: Conference Series **467** (2013) 012001 doi:10.1088/1742-6596/467/1/012001

oscillating continuously, and interfering so as to produce short pulses of light. During the dark time between pulses, the laser modes cancel each other by mutual interference.

Today, laser frequency combs are commercially available as ultra-precise spectroscopic measuring tools. A compact fiber-based frequency comb for space applications will be launched with a Texus sounding rocket in the Fall of 2013. The frequency range of laser combs is being extended to new regions from the THz regime to the extreme UV. Quite a few laboratories are exploring new approaches to create highly miniaturized comb generators, based on cascaded four-wave mixing in toroidal micro-resonators [3,4].

The original motivation for the invention of the frequency comb technique in our laboratory has been precision laser spectroscopy of the simple hydrogen atom, which permits unique confrontations between experiment and quantum electro-dynamic theory. The frequency of the particularly sharp 1S-2S two-photon resonance has recently been measured with an uncertainty of 4.2×10^{-15} [5,6]. Increasing control over systematic uncertainties may make it feasible for an atomic hydrogen optical clock to compete with ion clocks or optical lattice clocks, so that the second as unit of time could be redefined in terms of the simplest atom. Laser spectroscopy of additional transitions in hydrogen could help to resolve the "proton size puzzle" which has been revealed by laser spectroscopy of the 2S-2P Lamb shift in muonic hydrogen [7,8]. In the near future, laser spectroscopy of magnetically trapped antihydrogen should become feasible, so that one can search for conceivable differences between matter and antimatter.

Other applications of laser frequency combs have been evolving rapidly. Astronomers can use laser combs as calibration sources [9] for highly resolving spectrographs to search for extra-solar planets or to observe the postulated accelerating expansion of space in the universe. The control of the electric field of ultrafast laser pulses with frequency comb techniques has become [10] a key for attosecond science. Amongst the unforeseen applications of frequency combs, broad spectral bandwidth linear and nonlinear spectroscopy holds much promise: we can study complex molecular spectra in a highly multiplexed manner by using a hundred thousand or a million comb lines all at once.

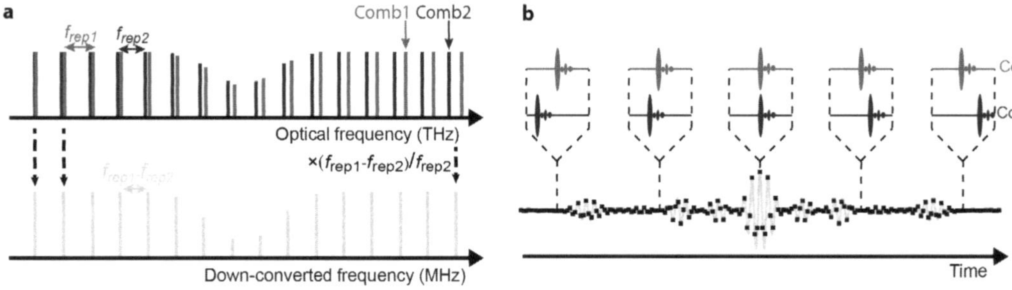

Figure 1. a) Frequency domain picture of the physical principle of dual-comb spectroscopy. Due to interference between pairs of optical comb lines, the optical spectrum is effectively mapped into a radio-frequency comb. The frequencies of the optical spectrum are down-converted to $\Delta f_{ceo} + n\ (f_{rep1} - f_{rep2})$, where Δf_{ceo} is the difference of carrier-envelope slippage frequencies of the two combs and n an integer. b) Time domain picture of the physical principle of dual-comb spectroscopy. The pulses of two combs with slightly different repetition periods interfere. The pulses from each comb overlap at linearly increasing time delays, mimicking a scanning interferometer.

Fourier Transform Spectroscopy [11] is a known method of highly multiplexed spectroscopy. This technique is in widespread use in many laboratories. The basic approach has not changed much since this technique emerged in the early seventies [12]. A Fourier spectrometer uses a (incoherent) light source, a moving-arm Michelson interferometer, and a single photo-detector connected to the data

Journal of Physics: Conference Series **467** (2013) 012001 doi:10.1088/1742-6596/467/1/012001

acquisition electronics of a computer. The autocorrelation function of the light, as filtered by an absorbing sample, the "interferogram", is Fourier-transformed to reveal the spectrum.

The incoherent light source can be replaced by a broadband laser frequency comb source of repetition frequency f_{rep1}, and the light reflected of a moving mirror can be replaced by that of a second laser frequency comb, tuned to a slightly different pulse repetition frequency f_{rep2}. This type of dual-comb spectroscopy offers several important advantages. First it uses coherent laser light rather than incoherent light. Secondly it does not require any moving parts. By simply dialling a knob, one can simulate a mirror moving at 10 km/s, the escape velocity from earth. In this way, 1/f noise in the detector signal can be greatly reduced, and a million-fold improvement in the acquisition speed of a spectrum has been demonstrated.

In the frequency domain (Fig.1a), dual-comb spectroscopy can be understood by considering the two frequency combs of slightly differing line spacing. Each pair of optical lines, one from each comb, produces a radio frequency beat note on the detector. These beat signals form a frequency comb in the radio frequency regime, spaced by f_{rep1}-f_{rep2}. Optical frequencies are thus converted into radio frequencies with a conversion factor $(f_{rep1}-f_{rep2})/f_{rep2}$.

In the time domain (Fig.1b), dual-comb spectroscopy can be described in the language of asynchronous optical sampling. A first pulse excites the molecules and a second pulse observes the free induction decay by interferometric sampling. As in a sampling oscilloscope, a periodic waveform, sampled asynchronouly, appears stretched in time by a factor $f_{rep2}/(f_{rep1}-f_{rep2})$, so that the signal can be recorded electronically and digitally processed.

Linear absorption dual-comb spectroscopy (Fig. 2a) requires two frequency comb sources, a single fast photodetector and a computer. It has been used [13-23] to demonstrate very short acquisition times on the order of microseconds, extreme sensitivity when used with cavity enhancement, and low to extreme resolution. Extreme accuracy is possible, the technique can reveal absorption (Fig. 2b) and dispersion, and it can be applied anywhere from THz frequencies to the VUV. By recording signals of some seconds, individual comb lines (Fig.3) are resolved [22]. A molecular spectrum is thus only sampled at discrete frequencies, but frequencies that can be known with the accuracy of an atomic clock. The comb spectrum can be scanned or interleaved for more complete coverage.

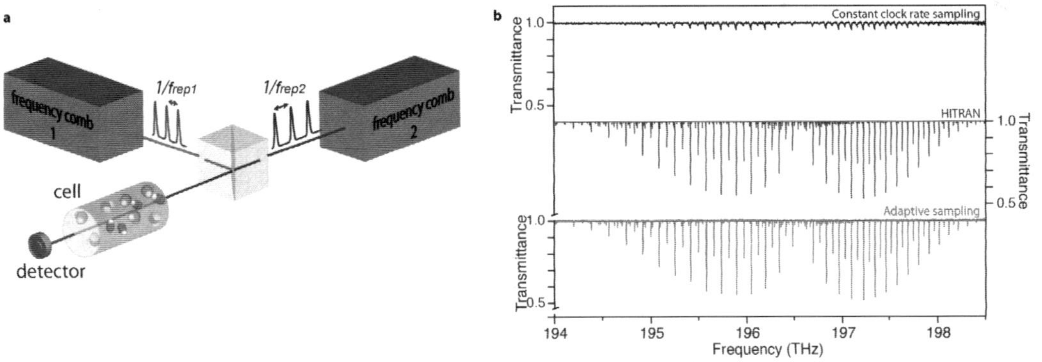

Figure 2. a) Dual-comb absorption spectroscopy. Two frequency combs, 1 and 2, have slightly repetition frequencies. Their beams are combined and are transmitted through the cell. They beat on a fast photodiode. b) Experimental dual-comb absorption spectra of $^{12}C_2H_2$ and $^{12}C^{13}CH_2$. Two experimental spectra are compared to a spectrum (blue) computed from the line parameters available in the 2008 HITRAN database. Both experimental spectra are measured with free-running fs lasers in identical conditions except for the interferometric real-time signal processing data acquisition clock. Two hundred consecutive spectra with an apodized resolution of 1.1 GHz, each measured within 467 μs, are averaged, resulting in a total measurement time of 93 ms and a total experimental time of 580 ms. When sampled at the constant clock rate of the digitizing data acquisition board, the spectrum

(black) is strongly distorted by the relative short-term fluctuations of the repetition frequency and carrier envelope phase slips of the fs lasers. This lowers the sensitivity and scrambles the spectral features. When compensating for the laser instabilities in real-time, the agreement of the experimental "adaptive" spectrum (red) with the computed spectrum is excellent and confirms the suitability of our technique for measurements with Doppler-limited resolution.

Despite its intriguing potential for the measurement of molecular absorption spectra spanning tens of nanometers within tens of microseconds at Doppler- limited resolution, the development of dual-comb spectroscopy is hindered by the extremely demanding requirements on the short-term stability of the laser combs. We have overcome this difficulty and experimentally demonstrated a straightforward concept of real-time dual-comb spectroscopy (Fig. 2b, [22,23]), which compensates for laser instabilities by electronic signal processing. It only uses free-running mode-locked lasers without any phase-lock electronics, a posteriori data-processing, or the need for expertise in frequency metrology.

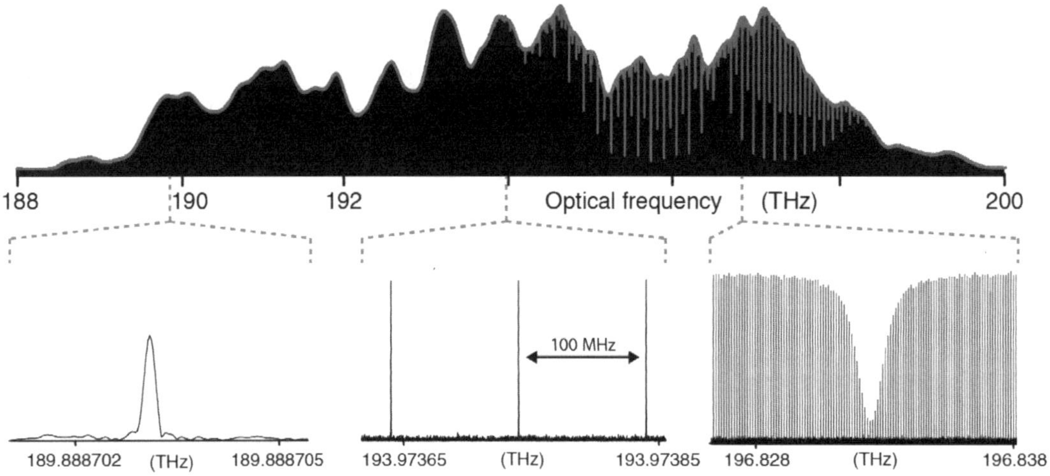

Figure 3. Near-infrared dual-comb spectrum with resolved comb lines. The spectrum across the entire domain of emission of Erbium femtosecond fiber lasers exhibits 120 000 well-resolved individual comb lines. Several acetylene absorption profiles shaping the discrete comb line intensities. The measurement time is 2.7 s and the optical instrumental linewidth is 200 kHz. In the lowest row, different degree of zooms in three different spectral regions well apart from these two frequencies isolate the individual comb lines and prove that the coherence between the two combs is maintained over the time of the measurement.

Since laser frequency combs involve intense ultrashort laser pulses, nonlinear interactions can be harnessed such as four wave mixing, intensity-dependent refractive index changes, light induced birefringence and dichroism, or other coherent transient phenomena including photon echoes.
Direct frequency comb Doppler-free spectroscopy of two-photon resonances has already been demonstrated in our first frequency comb experiments [24,25] with J. Eckstein and A. Ferguson at Stanford around 1976. As pointed out by E.V. Baklanov and V.P. Chebotaev [26], many pairs of comb lines with the same sum frequency contribute to the excitation of an atomic resonance, so that the excitation probability can be the same as for a resonantly tuned cw laser of the same average power. However, any observed resonance repeats with the line spacing of the comb, so that this type of direct frequency comb spectroscopy is only suitable for simple spectra composed of very few transitions.

Journal of Physics: Conference Series **467** (2013) 012001 doi:10.1088/1742-6596/467/1/012001

Recently, we have demonstrated [27] the new technique of dual-comb two-photon spectroscopy, which identifies each transition uniquely by the modulation imparted by the interfering excitations. It can combine sub-Doppler resolution with a free spectral range only limited by the spectral bandwidth of the laser frequency combs. Proof-of-principle demonstration has been achieved with the 5S-5D two-photon transitions of rubidium (Fig. 4).

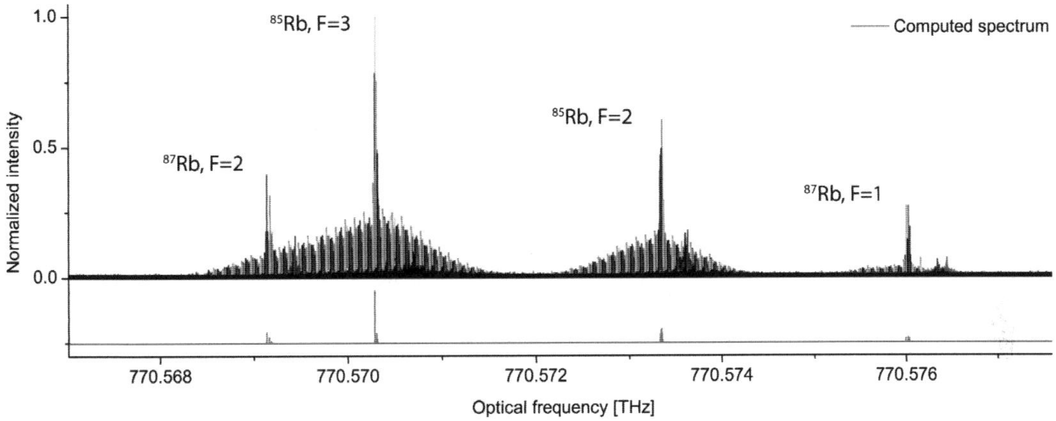

Figure 4. Two-photon excitation sub-Doppler dual-comb spectroscopy of the $5S_{1/2}$ -$5D_{5/2}$ transitions of natural rubidium. Doppler-free resonances emerging the Doppler-broadened resonances are clearly distinguishable. Thirteen individual spectra with resolved comb lines of a spacing of 100 MHz are interleaved, improving the resolution to 8 MHz. The measurement time for the full-interleaved spectrum is 52 s. F denotes the hyperfine quantum number in the ground state.

In other experiments [28,29], we harness coherent Raman effects. We have demonstrated [29] coherent anti-Stokes Raman spectroscopy and spectro-imaging of liquid samples with two laser frequency combs. All spectral elements are simultaneously measured on a single photodetector within a short time on the microsecond scale. The spectral bandwidth is determined by the span of the ultrashort pulse lasers. The resolution is only limited by the physical width of the molecular bands. With samples of liquid solvents, we demonstrate (Fig. 5) high resolution (4 cm^{-1}) Raman spectra spanning more than 1200 cm^{-1} recorded within tens of microseconds and these capabilities have been extended to hyperspectral imaging with an acquisition rate of 50 pixels/s. Real-time multiplex accessing of hyperspectral images may expand the range of applications of nonlinear microscopy.

As laser frequency comb techniques and laser technology continue to evolve, new applications come within reach. Curiosity-driven research with new instruments will lead to surprising discoveries. Let us end with a quote from Charles Townes („How the laser happened") [30]: „There is much that we don't understand; in many cases, we don't understand that we don't. And the really surprising discoveries will probably depend primarily on individuals, not teams or committees." The opportunities for individuals to make important and surprising contributions is perhaps the most important secret of success of our field of laser spectroscopy.

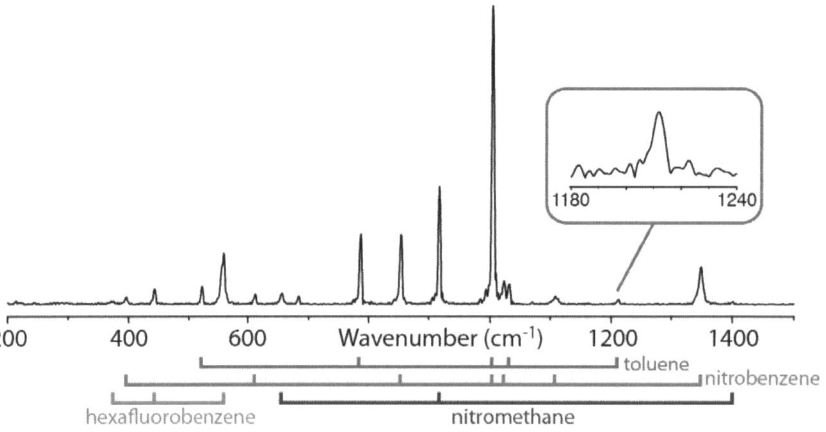

Figure 5. Dual-comb coherent anti-Stokes Raman high-resolution spectroscopy of a mixture of liquid chemicals. The spectrum measured within 295.5 μs at 4 cm^{-1} apodized resolution. The number of individual spectral elements (defined as spectral span divided by resolution) is 300.

References

[1] Hänsch T W 1972, *Appl. Opt.* **11** 895-8

[2] Hänsch T W 2006, *Rev. Mod. Phys.* **78**, 1297-1309

[3] Kippenberg T J, Holzwarth R, Diddams S A 2011, *Science* **332**, 555–9

[4] Wang C Y, Herr T, Del'Haye P, Schliesser A, Hofer J, Holzwarth R, Hänsch T W, Picqué N, Kippenberg T J 2013, *Nature Communications* **4**, 1345

[5] Parthey C G, Matveev A, Alnis J, Bernhardt B, Beyer A, Holzwarth R, Maistrou A, Pohl R, Predehl K, Udem T, Wilken T, Kolachevsky N, Abgrall M, Rovera D, Salomon C, Laurent P, Hänsch T W 2011, *Physical Review Letters* **107**, 203001

[6] Matveev A, Parthey C G, Predehl K, Alnis J, Beyer A, Holzwarth R, Udem T, Wilken T, Kolachevsky N, Abgrall M, Rovera D, Salomon C, Laurent P, Grosche G, Terra O, Legero T, Schnatz H, Weyers S, Altschul B, Hänsch T W 2013, *Physical Review Letters* **110**, 230801

[7] Pohl R, Antognini A, Nez F, Amaro F D, Biraben F, Cardoso J M R, Covita D S, Dax A, Dhawan S, Fernandes L M P, Giesen A, Graf T, Hänsch T W, Indelicato P, Julien L, Kao C-Y, Knowles P, Lopes J A M, Le Bigot E-O, Liu Y-W, Ludhova L, Monteiro C M B, Mulhauser F, Nebel T, Rabinowitz P, dos Santos J M F, Schaller L A, Schuhmann K, Schwob C, Taqqu D, Veloso J F C A, Kottmann F 2010, *Nature* **466**, 213–216

[8] Antognini A, Nez F, Schuhmann K, Amaro F D, Biraben F, Cardoso J M R, Covita D S, Dax A, Dhawan S, Diepold M, Fernandes L M P, Giesen A, Gouvea A L, Graf T, Hänsch T W, Indelicato P, Julien L, Kao C Y, Knowles P, Kottmann F, Le Bigot E-O, Liu Y-W, Lopes J A M, Ludhova L, Monteiro C M B, Mulhauser F, Nebel T, Rabinowitz P, dos Santos J M F, Schaller L A, Schwob C, Taqqu D, Veloso J F C A, Vogelsang J, Pohl R 2013, *Science* **339**, 417-420

[9] Wilken T, Lo Curto G, Probst R A, Steinmetz T, Manescau A, Pasquini L, González Hernández J I, Rebolo R, Hänsch T W, Udem T, Holzwarth R 2012, *Nature* **485**, 611-614

[10] Baltuska A, Udem T, Uiberacker M, Hentschel M, Goulielmakis E, Gohle C, Holzwarth R, Yakovlev V S, Scrinzi A, Hänsch T W, Krausz F 2003, *Nature* **421**, 611-615

[11] Griffiths P R, De Haseth J A 2007, Fourier Transform Infrared Spectroscopy. John Wiley & Sons Inc..

[12] Connes J, Delouis H, Connes P, Guelachvili G, Maillard J-P, Michel G 1970, *Nouv. Rev. Opt. Appl.* **1**, 3-22

[13] Bernhardt B, Ozawa A, Jacquet P, Jacquey M, Kobayashi Y, Udem T, Holzwarth R, Guelachvili G, Hänsch T W, Picqué N 2010, *Nat. Photon.* **4**, 55-57

[14] Coddington I, Swann W C, Newbury N R 2008, *Phys. Rev. Lett.* **100**, 013902

[15] Bernhardt B, Sorokin E, Jacquet P, Thon R, Becker T, Sorokina I T, Picqué N, Hänsch T W 2010, *Appl Phys B-Lasers O* **100**, 3-8.

[16] Schliesser A, Picqué N, Hänsch T W 2012, *Nat. Photon.* **6**, 440-449

[17] Keilmann F, Gohle C, Holzwarth R 2004, *Opt. Lett.* **29**, 1542-1544

[18] Schliesser A, Brehm M, Keilmann F, van der Weide D W 2005, *Opt. Express* **13**, 9029-9038

[19] Yasui T, Saneyoshi E, Araki T 2005, *Appl. Phys. Lett.* **87**, 061101

[20] Coddington I, Swann W C, Newbury N R 2010, *Phys. Rev. A* **82**, 043817

[21] Giaccari P 2008, *Opt. Express* **16**, 4347-4365

[22] Ideguchi T, Poisson A, Guelachvili G, Hänsch T W, Picqué N 2012, *Opt. Lett.* **37**, 4847-4849

[23] Ideguchi T, Poisson A, Guelachvili G, Picqué N, Hänsch T W 2012, arXiv:1201.4177

[24] Teets R, Eckstein J, and Hänsch T W1977, *Phys. Rev. Lett.* **38**, 760–764

[25] Eckstein J N, Ferguson A I, Hänsch T W 1978, *Phys. Rev. Lett.* **40**, 847–850

[26] Baklanov Y V, Chebotayev V P 1977, Appl. Phys. **12**, 97- 99

[27] Hipke A, Meek S A, Guelachvili G, Hänsch T W, Picqué N 2013, in CLEO: 2013, OSA Technical Digest (Optical Society of America, 2013), paper CTh5C.8.

[28] Ideguchi T, Bernhardt B, Guelachvili G, Hänsch T W, Picqué N 2012, *Optics letters* **37**, 4498-4500

[29] Ideguchi T, Holzner S, Bernhardt B, Guelachvili G, Hänsch T W, Picqué N 2013, arXiv:1302.2414

[30] Townes C H 2000, How the Laser Happened: Adventures of a Scientist, Oxford University Press.

Journal of Physics: Conference Series **467** (2013) 012002 doi:10.1088/1742-6596/467/1/012002

Ultra-stable long distance optical frequency distribution using the Internet fiber network and application to high-precision molecular spectroscopy.

O Lopez[1], B Chanteau[1], A. Bercy[1], D Nicolodi[2], W Zhang[2], B. Argence[1], M Abgrall[2], A Haboucha[2], A Kanj[2], D Rovera[2], J Achkar[2], P-E Pottie[2], B Darquié[1], C Chardonnet[1], Y Le Coq[2], G Santarelli[2,3] and A Amy-Klein[1]

[1]Laboratoire de Physique des Lasers, Université Paris 13, Paris Cité Sorbonne, CNRS, 99 Av. Jean-Baptiste Clément, 93430 Villetaneuse, France

[2]LNE-SYRTE, Observatoire de Paris, CNRS, UPMC, 61 Av. de l'Observatoire, 75014 Paris, France

[3]Laboratoire Photonique, Numérique et Nanosciences, UMR 5298 Université de Bordeaux 1, Institut d'Optique and CNRS, 351 Cours de la Libération, 33405 Talence, France

amy@univ-paris13.fr

Abstract. We report an optical link of 540 km for ultrastable frequency distribution over the Internet fiber network. The phase-noise compensated link shows a fractional frequency instability in full bandwidth of 3×10^{-14} at one second measurement time and 2×10^{-18} at 30 000 s. This work is a significant step towards a sustainable wide area ultrastable optical frequency distribution and comparison network. Time transfer was demonstrated simultaneously on the same link and led to an absolute time accuracy (250 ps) and long–term timing stability (20 ps) which outperform the conventional satellite transfer methods by one order of magnitude. Current development addresses the question of multiple users distribution in the same metropolitan area. We demonstrate on-line extraction and first results show frequency stability at the same level as with conventional link. We also report an application to coherent frequency transfer to the mid-infrared. We demonstrate the frequency stabilisation of a mid-infrared laser to the near-infrared frequency reference transferred through the optical link. Fractional stability better than 4×10^{-14} at 1 s averaging time was obtained, opening the way to ultrahigh resolution spectroscopy of molecular rovibrational transitions.

1. The challenge of ultrastable frequency dissemination

Frequency metrology has developed considerably over the past ten years and has benefited from scientific advances in the areas of atom laser cooling and frequency comparison with femtosecond laser combs. Today cold atom microwave frequency standards reach routinely a fractional accuracy in the low 10^{-16} and trapped ion or neutral lattice optical clocks have already demonstrated uncertainty of parts in the 10^{17} or better (see for instance [1-5]). This outstanding performance makes them ideal tools for fundamental physics tests of the validity of general relativity on earth and in space. Among them, the comparison of different types of clocks is used to detect possible variations in time of

Journal of Physics: Conference Series **467** (2013) 012002 doi:10.1088/1742-6596/467/1/012002

universal constants of physics. More generally, accurate frequency transfer is essential for the underpinning of the uncertainty of almost every type of precision measurement.

Until recently the conventional means for remote frequency transfer was based on the processing of satellite radio-frequency signals, the Global Navigation Satellite System, or dedicated satellite transfer. Optical fiber links have brought the potential to transfer frequency with much better accuracy and stability thanks to an active compensation of the phase noise added by the fluctuations of the optical length. Direct optical frequency transfer has been developed in order to take full advantage of the optical fiber medium. The link-induced phase noise can indeed be detected with much higher resolution than with a microwave signal frequency and heterodyne detection enables to minimize the effect of link attenuation. In the last five years several experiments all around the world have explored the limits of this method, paving the way to ultrastable optical networks of new generation [6-11]. The latest achievement is an outstanding phase-stabilized 920 km fiber link connecting two German laboratories [10]. A key-question is the capability to extend ultra-stable frequency standard distribution to a larger scale and possibly to any users. In the last four years we have demonstrated optical links using commercial telecommunication fiber networks thanks to wavelength division multiplexing [11-13]. Switching from a dedicated to a public fiber network with data traffic is a major breakthrough for a possible generalization of the concept of optical fiber link frequency distribution. In this paper we demonstrate a robust long distance 540-km phase-stabilized optical link over public telecommunication network for the dissemination of ultra-stable frequency around 1.55 μm. We show that this link can simultaneously been used to transfer accurate timing signal. Then we discuss possible architecture of metropolitan signal distribution and give first results of on-line extraction. We finally show an application of such an optical link to coherent frequency transfer to the mid-Infrared for high resolution molecular spectroscopy.

2. Optical link over public network

We first give the basic principle of an optical link. The frequency reference is an ultra-stable laser emitting around 1.55 μm, which is feeding the optical fiber and transferred to the remote lab. Since the propagation is disturbed by phase noise, due to the thermal and acoustical instabilities of the optical length of the fiber, a correction loop is implemented using the so-called round-trip method. Part of the transferred signal is sent back to local lab and the return signal is compared with the local reference signal to get the instability of the link: the phase difference gives the error signal for the correction loop. Noise compensation is usually applied to an acousto-optic modulator (AOM). The performance of such a link is evaluated by measuring the end-to-end beatnote, which is possible only when the local and remote ends are at the same place. Thus we will show below optical links made of two parallel fibers.

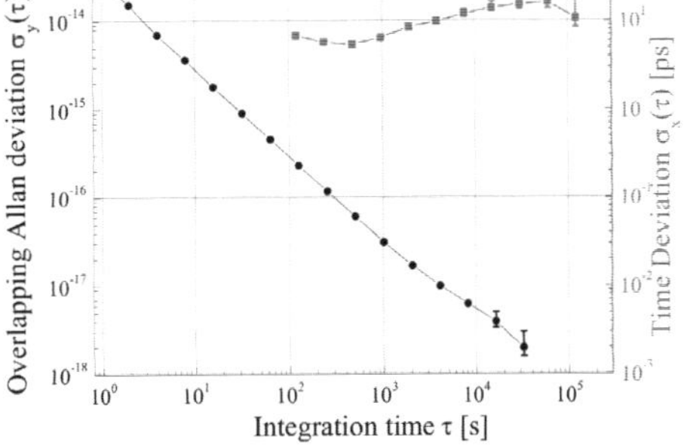

Figure 1. Time (red squares) and frequency (black circles) stability for a 540-km optical link over public network

A multiplexed optical link uses this noise compensation on installed telecommunication fibers [11-13]. The ultrastable frequency reference and the Internet traffic propagate simultaneously in the same fiber but on different frequency channels. Hence, instead of using a dark fiber, we are using a dark channel of existing fibers, thus dramatically reducing the need and cost of fiber infrastructures.

The key point to understand the architecture of a multiplexed optical link is to keep in mind that stable optical signal distribution requires a full bidirectional operation: the optical signal must circulate back and forth on the same fiber for noise compensation. This means that in a single fiber there are two counter-propagating optical

Journal of Physics: Conference Series **467** (2013) 012002 doi:10.1088/1742-6596/467/1/012002

signals whereas standard telecommunication technique enables only one way propagation. Therefore, each telecommunication equipment, such as amplifier or repeater, must be bypassed by the ultrastable signal. For that purpose, we use bidirectional optical add-drop multiplexers (OADM). Such three-port component can insert or extract a specified wavelength from the other wavelengths, with a very good isolation and low losses.

With this technique we demonstrated an optical link of 540-km from LPL to Reims and back to LPL [11], using the fiber network of the French national network for education and research RENATER. This link is composed of different fiber spans and includes 450-km with data traffic on neighbor frequency channels, 0.8 nm and 1.6 nm away from the ultrastable signal frequency. Sixteen OADM are necessary to insert and extract the ultrastable signal from the Internet fibers. We used six bidirectional Erbium-doped fiber amplifiers (EDFA), with a total amplification of about 100 dB, to partly compensate the total end-to-end attenuation of 165 dB due to fiber losses, OADMs and the large number of connectors. Attenuation cannot be compensated due to the parasitic reflections along the link which prevent to increase the gain of bidirectional amplifiers in order to avoid oscillations.

Figure 1 shows the fractional frequency instability (overlapping Allan deviation, black circles) of the 540 km link, measured with a Π-type frequency counter. The Allan deviation is 3×10^{-14} at 1 s averaging time and scales down as $1/\tau$ from 1 s to 30 000 s reaching about 2×10^{-18}. With a measurement bandwidth of 10 Hz, the Allan deviation is about 8 times smaller and reaches 5×10^{-15} at 1 s and 3×10^{-19} at 30 000 s [10]. Moreover the average frequency offset between input and output frequency is 6×10^{-20} compatible with statistical error bars of 3×10^{-19}. This stability is similar to the one obtained with the 920-km German optical link [10], that exhibits a fractional instability (overlapping Allan deviation, Π-data) of 4×10^{-14} at 1 s reaching 5×10^{-18} at 10^4 s. With Λ-data and calculating modified Allan deviation, this stability reaches 5×10^{-15} at 1 s and 10^{-18} in less than 10^3 s.

For many applications, concerning for instance high resolution radio-astronomy and particle physics, frequency transfer is not sufficient and accurate timing is required to precisely synchronize distant experiments. Time transfer consists in comparing the epochs of two distant clocks and not only their frequencies. It demands an accurate calibration of the propagation delay. Using the same 540-km optical link we have demonstrated two-way time transfer [14] simultaneously to the frequency transfer [15]. The ultra-stable laser is carrying both the frequency and timing signals through the fiber. Timestamps signals are consisting in spread spectrum pseudorandom modulation codes and are provided by a pair of two-way satellite time transfer modems, one for each direction. They are encoded on the optical carrier by electro-optic phase modulators. The timing stability (red squares) is below 20 ps for integration time up to 10^5 s (see Figure 1). The time delay variation remains below 50 ps when the fiber length is scaled from 10 m up to 540 km. This method is very robust, with an overall conservative timing error of about 250 ps determined by the measured time delay over several weeks of measurement. These results outperform the satellite techniques by one order of magnitude. They are easily scalable to a wide area network and open the way to a novel approach of precise remote synchronization with the potential to realize even better accuracy.

3. On-line-extraction

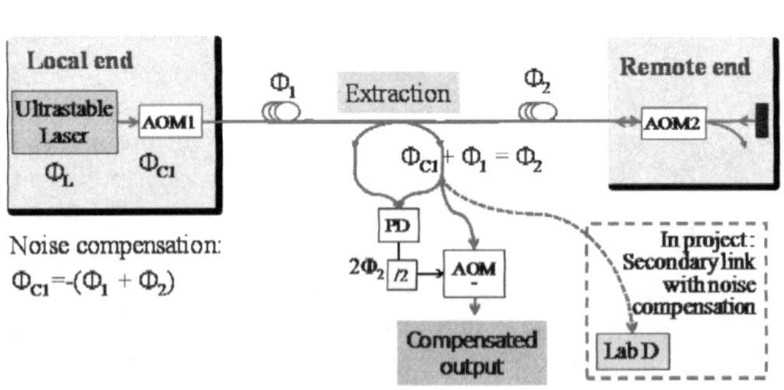

Figure 2. On-line extraction scheme on a main compensated link. AOM : acousto-optic modulator, PD : photodiode

The extension of the ultrastable frequency signal dissemination to a large number of laboratories asks the question of the network architecture. For long-distance transfer, cascaded optical links have already been proposed and demonstrated [7, 13]. Here we address the question of multiple users distribution. In that case, a point-to point network needs a lot of fibers and is not the optimum solution. On-line extraction, as first proposed three years ago by G. Grosche [16], enables a more flexible distribution: the signal is extracted from the main link and either directly distributed to a lab or fed to another secondary link. Figure 2 displays a schematic scheme of the link architecture. The noise compensation of the main link leads to an overcorrection at the extraction point. In order to compensate this over-correction, both extracted signals are recombined to get the relative phase. After division by two, this signal is driven an AOM in order to correct for the over-noise. With that scheme, the phase noise due to the main link is compensated.

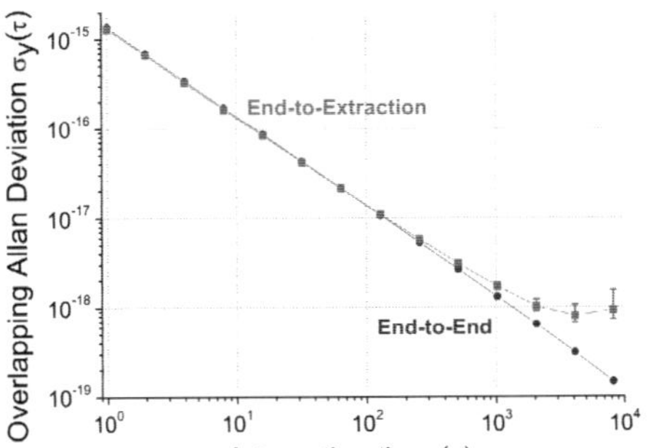

Figure 3. Fractional frequency stability for a 92-km link on dedicated fibers (blue circles) with noise compensation, and for on-line extraction after 86-km (red squares).

This set-up has been tested on an optical link of 92-km dedicated fiber from the telecommunication network in the Parisian metropolitan area. Both ends are at LNE-SYRTE, together with the extraction, which occurred after 86-km of the main link. Figure 3 displays the end-to-end frequency stability of the main link (blue circles) with noise compensation. On the same graph is given the extracted signal stability (red squares), which is the stability of the beatnote between the extracted signal and the local signal. It is below 2×10^{-18} at 10^3 s integration time thus providing the stability performance at the same level than with a simple link. However the stability degrades around 10^3 s due to temperature fluctuations of the extraction set-up. There are indeed still some uncompensated sections of fiber at the extraction end, mainly between the AOM and the user.

To improve the noise compensation, we are now implementing an upgraded extraction scheme. A short secondary link is implemented just after the extraction. Its noise compensation is added to the previous compensation in order to compensate both the noise of the main link and the local noise. This will give the possibility to distribute an ultrastable signal in a wide metropolitan area.

4. Coherent frequency transfer to the mid-infrared

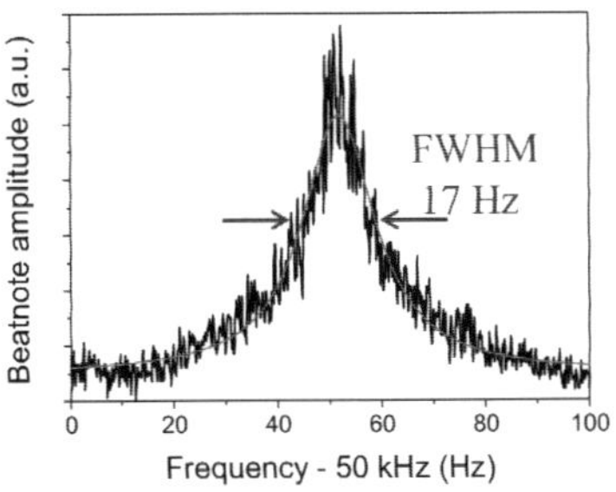

Figure 4. Down-converted beatnote of two independent CO_2 lasers, one stabilised on an OsO_4 resonance line, the other phase-locked to a comb repetition rate high-harmonic; red line is a Lorentzian fit of linewidth 17 Hz

Such an optical link enables any physics laboratory to access an ultrastable reference signal and opens the way to a wide range of new experiments. Among them, ultrastable and accurate laser spectroscopy gives the possibility to test fundamental physics with atoms and molecules. As a first application, we demonstrate the phase-locking of a mid-infrared (MIR) laser located at LPL to the remote 1.54 μm ultra-stable laser from LNE-SYRTE, which is measured against a set of primary standards [1, 17] using an optical frequency comb [18]. Once transferred, the signal is used to phase-lock a second optical frequency comb. Using sum-frequency generation in a non-linear crystal [19], the MIR laser is in turn locked to a high-harmonic of the comb repetition rate. With this set-up, we stabilized a CO_2 laser around 10 μm and obtained a linewidth below 17 Hz (see Figure 4), at the state-of-the art. Stability below

4×10^{-14} at 1 s was demonstrated, and we expect it to be in the 10^{-15} range, at the level of the near-infrared frequency reference [20]. The 3×10^{-16} accuracy of the LNE-SYRTE Cs fountains is potentially within reach. Tunability is also possible by scanning the near-infrared frequency referencing the comb. Extension to Quantum Cascade Laser frequency stabilization is under progress. Such a stabilized laser gives the possibility to perform ultra-high resolution spectroscopy of molecules and test fundamental symmetries such as parity [21].

5. Perspectives

To conclude, we have established an optical link of 540-km which uses an optical telecommunication network simultaneously carrying Internet data. We demonstrate joint frequency and time transfer over this link with frequency instability integrating down to 2×10^{-19} after 30000 s. This result is a first step towards an extension of the fiber network in France and Europe using several cascade optical links connected by stations.

These development opens the way to very fascinating applications of optical links. Transcontinental optical links will give the opportunity to compare the best primary and optical clocks in different countries. We plan also to take advantage of fiber links to evaluate the performance of satellite links, such as the microwave link of the ACES space mission. At LPL, we are currently taking advantage of the ultrastable frequency transfer to test fundamental symmetries with high resolution spectroscopy of molecules. Finally optical links open the way to numerous new high precision experiments, as for instance ultra-precise atomic and molecular spectroscopy, measurement of fundamental constants or fundamental tests of physics like the search for space-time variations of fundamental constants. And the number of these applications will grow with the extension of the fiber network.

Acknowledgments

This work would not have been possible without the support of the GIP RENATER. The authors acknowledge financial support from CNRS, Agence Nationale de la Recherche (ANR BLANC 2011-BS04-009-01), Université Paris 13 and IFRAF-Conseil Régional Ile-de-France.

References

[1] Guena J, Abgrall M, Rovera D, Laurent P, Chupin B, Lours M, Santarelli G, Rosenbusch P, Tobar M E, Li R, Gibble K, Clairon A and Bize S 2012 Progress in atomic fountains at LNE-SYRTE *IEEE Trans. on Ultras. Ferro. and Frequ. Contr.* **59** 391

[2] Chou C, Hume D B, Thorpe M J, Wineland D J and Rosenband T 2011 Quantum coherence between two atoms beyond $Q=10^{15}$ *Phys. Rev. Lett.* **106** 160801

[3] Katori H 2011 Optical lattice clocks and quantum metrology Nature Photon. **5** 203 and references therein.

[4] Le Targat R., Lorini L, Le Coq Y, Zawada M, Guéna J, Abgrall M, Gurov M, Rosenbusch P, Rovera D G, Nagórny B, Gartman R, Westergaard P. G., Tobar M E, M. Lours M, G. Santarelli G, A. Clairon A, S. Bize S, Laurent P, Lemonde P and and J. Lodewyck 2013 Experimental realization of an optical second with strontium lattice clocks *Nature Communications* **4** 2109

[5] Hinkley N, Sherman J A, Phillips N B, Schioppo M, Lemke N D, Beloy K, Pizzocaro M, Oates C W and Ludlow A D 2013 An atomic clock with 10^{-18} instability *arXiv:1305.5869v1*

[6] Newbury R, Williams P A and Swann W C 2007 Coherent transfer of an optical carrier over 251 km *Opt. Lett.* **32** 3056

[7] Jiang H, Kéfélian F, Crane S, Lopez O, Lours M, Millo J, Holleville D, Lemonde P, Chardonnet C, Amy-Klein A and Santarelli G 2008 Long-distance frequency transfer over an urban fiber link using optical phase stabilization *J. Opt. Soc. Am. B* **25** 2029

[8] Grosche G, Terra O, Predehl K, Holzwarth R, Lipphardt B, Vogt F, Sterr U and Schnatz H 2009 Optical frequency transfer via 146 km fiber link with 10^{-19} relative accuracy *Opt. Lett.* **34**

2270

[9] Fujieda M, Kumagai M, Nagano S, Yamaguchi A, Hachisu H and Ido T 2010 All-optical link for direct comparison of distant optical clocks *Opt. Express* **19** 16498

[10] Predehl K, Grosche G, Raupach S M F, Droste S, Terra O, Alnis J, Legero Th, Hänsch T W, Udem Th, Holzwarth R and Schnatz H 2012 A 920-Kilometer optical fiber link for frequency metrology at the 19th decimal place *Science* **336** 441

[11] Lopez O, Haboucha A, Chanteau B, Chardonnet C, Amy-Klein A and Santarelli G 2012 Ultra-stable long distance optical frequency distribution using the Internet fiber network *Optics Express* **20** 23518

[12] Kéfélian F, Lopez O, Jiang H, Chardonnet Ch, Amy-Klein A and Santarelli G 2009 High-resolution optical frequency dissemination on a telecommunication network with data traffic *Opt. Lett.* **34** 1573

[13] Lopez O, Haboucha A, Kéfélian F, Jiang H, Chanteau B, Roncin V, Chardonnet C, Amy-Klein A and Santarelli G 2010 Cascaded multiplexed optical link on a telecommunication network for frequency dissemination *Optics Express* **18** 16849

[14] Rost M, Piester D, Yang W, Feldmann T, Wübbena T and Bauch A 2012 Time transfer through optical fibers over a distance of 73 km with an uncertainty below 100 ps *Metrologia* **49** 772

[15] Lopez O, Kanj A, Pottie P E, Rovera D, Achkar J, Chardonnet C, Amy-Klein A, Santarelli G 2013 Simultaneous remote transfer of accurate timing and optical frequency over a public fiber network *Applied Physics B* **110** 3

[16] Grosche G, Physikalisch-Technische Bundesanstalt, Braunschweig, Germany; patent application DE 10.2008.062.139, Method for making available a reference frequency (personal communication, 2010).

[17] Chambon D, Bize S, Lours M, Narbonneau F, Marion H, Clairon A, Santarelli G, Luiten A and Tobar M 2005 Design and realization of a flywheel oscillator for advanced time and frequency metrology *Review of Scientific Instruments* **76** 094704

[18] Zhang W, Lours M, Fischer M, Holzwarth R, Santarelli G and Le Coq Y 2012 Characterizing a fiber-based frequency comb with electro-optic modulator *IEEE Transactions on Ultrasonics, Ferroelectrics, and Frequency Control* **59** 432

[19] Amy-Klein A, Goncharov A, Guinet M, Daussy C, Lopez O, Shelkovnikov A and Chardonnet C 2005 Absolute frequency measurement of a SF_6 two-photon line by use of a femtosecond optical comb and sum-frequency generation *Optics Letters* **30** 3320

[20] Chanteau B, Lopez O, Zhang W, Nicolodi D, Argence B, Auguste F, Abgrall M, Chardonnet C, Santarelli G, Darquié B, Le Coq Y and Amy-Klein A 2013 Mid-infrared laser phase-locking to a remote near-infrared frequency reference for high precision molecular spectroscopy *New Journal of Physics* **15** 073003

[21] Darquié B, Stoeffler C, Shelkovnikov A, Daussy C, Amy-Klein A, Chardonnet C, Zrig S, Guy L, Crassous J, Soulard P, Asselin P, Huet T R, Schwerdtfeger P, Bast R and Saue T 2010 Progress toward a first observation of parity violation in chiral molecules by high-resolution laser spectroscopy *Chirality* **22** 870

Precision Spectroscopy of Atomic Hydrogen

A. Beyer[1], Ch. G. Parthey[1], N. Kolachevsky[1], J. Alnis[1], K. Khabarova[1], R. Pohl[1], E. Peters[1], D. C. Yost[1], A. Matveev[1], K. Predehl[1], S. Droste[1], T. Wilken[1], R. Holzwarth[1], T. W. Hänsch[1], M. Abgrall[2], D. Rovera[2], Ch. Salomon[3], Ph. Laurent[2] and Th. Udem[1]

[1] Max-Planck Institut für Quantenoptik, Hans-Kopfermann Str.1, 85748 Garching, Germany
[2] LNE-SYRTE, Observatoire de Paris, Paris, France
[3] Lab. Kastler-Brossel, CNRS, UPMC, departement de physique de l'ENS, 75231 Paris, France

E-mail: thomas.udem@mpq.mpg.de

Abstract. Precise determinations of transition frequencies of simple atomic systems are required for a number of fundamental applications such as tests of quantum electrodynamics (QED), the determination of fundamental constants and nuclear charge radii. The sharpest transition in atomic hydrogen occurs between the metastable 2S state and the 1S ground state. Its transition frequency has now been measured with almost 15 digits accuracy using an optical frequency comb and a cesium atomic clock as a reference [1]. A recent measurement of the $2S - 2P_{3/2}$ transition frequency in muonic hydrogen is in significant contradiction to the hydrogen data if QED calculations are assumed to be correct [2, 3]. We hope to contribute to this so-called "proton size puzzle" by providing additional experimental input from hydrogen spectroscopy.

1. Introduction

Atomic hydrogen has been the Rosetta Stone for quantum physics. Through the successive refinement of its theoretical description, starting from the phenomenological description of J.J Balmer, it provided a concise argument for the wave nature of matter. N. Bohr's quantization requirement, that the electrons angular momentum must be an integer multiple of $\hbar$, can be understood such that the corresponding de Broglie wave forms a standing wave. E. Schrödinger was the first to formulate a wave equation for matter wave. With his Schrödinger equation, he put the emerging field quantum mechanics on solid grounds, while P.A.M. Dirac figured out how to include relativity. These refinements were introduced to keep up with increasing experimental accuracy, and to fix the resulting inconsistencies, but continuously complicated the theoretical picture. W.E. Lamb and R.C. Retherford discovered a discrepancy with the Dirac theory in 1947. It turned out that this was due to the fact that quantum mechanics had not yet been applied in the most rigourous way. In particular, the effects of the quantum vacuum, both the electronic and photonic, had been neglected up to then. Their inclusion led to the development of Quantum Electrodynamics (QED) by E.A. Ühling, H.A. Bethe, R.P. Feynman and others. Unlike any of the predecessors, QED survived about six orders of magnitude improvement of the experimental accuracy and is still the current description today. Of course several discrepancies between experiment and theory have shown up in the last 65 years, but thus far they could all traced back to errors in the computation of QED terms, neglected higher order terms or to underestimated experimental errors. In that sense QED is probably the most successful theory in all of physics. As such it has served as a template for all subsequent quantum field theories.

A serious conceptual drawback of this development is not only that our description of nature has become more complicated but also that a parameter was introduced, the dimensionless fine structure constant $\alpha \approx 1/137$, in order to account for the observed fine structure of hydrogen spectral lines. Since we do not know any way to calculate fundamental constants, such as α, from first principles, the additional constant alpha reduces the predictive power of theory. In practice though, there are even more parameters that enter. The electron to proton mass ratio m_e/m_p and proton charge radius r_p take into account the finite mass and size of the nucleus respectively. Finally, most transition frequencies in hydrogen are measured in SI units, so that the Rydberg constant R_∞ appears as a multiplicative unit converter from atomic units to SI units. In QED the hydrogen energy levels can be expressed, like any other QED result, as a power series in α:

$$E = R_\infty \left(-\frac{1}{n^2} + A_{20}\alpha^2 + A_{30}\alpha^3 + A_{31}\alpha^3 ln(\alpha) + A_{40}\alpha^4 + \ldots + \frac{16\pi^2 m_e^2 c^2 \alpha^2}{3n^3 h^2} r_p^2 \right) \quad (1)$$

In leading order this reproduces the result of N. Bohr and E. Schrödinger. The dependence of m_e/m_p, as well as on the other quantum numbers, is hidden in the coefficients A_{mn} of which a complete collection of all the known ones are listed in [4]. The last term in eqn.(1) describes the effect of the finite size of the nucleus in terms of the r.m.s proton charge radius r_p for S-states. For hydrogen this term contributes very little to the total energy[1], but adds the largest part to the uncertainty, if the value of the proton charge radius is taken from elastic electron proton scattering (see for example table 2 of [5]). In muonic hydrogen the electron is replaced by a muon which possesses a ≈ 200 times larger mass. The finite size term increases by more than 4 orders relative to the other terms and by about 7 orders of magnitude in absolute terms, as the Rydberg constant is proportional to the orbiting lepton mass as well. Therefore the measurement of transition frequencies in muonic hydrogen provides a much more sensitive method to determine the proton charge radius compared to regular hydrogen.

2. Determining the Parameters

The theoretical description of the hydrogen energy levels uses four parameters: The fine structure constant α, the electron to proton mass ratio m_e/m_p, the proton charge radius r_p and the Rydberg constant R_∞. To determine their values one could use four measured transition frequencies in atomic hydrogen. However, other branches of physics may provide more precise values for some of them. This is the case for the fine structure constant which can be determined with a relative uncertainty of 3.7 parts in 10^{10} by comparing the measured value of the electron g-factor with the corresponding QED expression [6]. In contrast to eqn.(1) it is linear in α, so that the hydrogen energy levels are about 137 times less sensitive for the determination of α than the electron g-factor.

The other parameter that is advantageously taken from another experiment, given the accuracy of currently available data, is the electron to proton mass ratio. Similar to the fine structure constant, the hydrogen energy levels are not sensitive to this quantity in first-order. On the other hand, mass ratios of particles that can simultaneously be held in a Penning trap can be determined with relative uncertainties in the 10^{-10} range by comparing their cyclotron frequencies [7]. In this way the electron proton mass ratio has been determined [8] with an uncertainty lower than can be extracted using eqn.(1). This leaves us with two parameters r_p and R_∞ to be determined from two measured hydrogen transition frequencies.

[1] For that reason the additional parameters that appear there can be taken with a few digits accuracy from any other experiment.

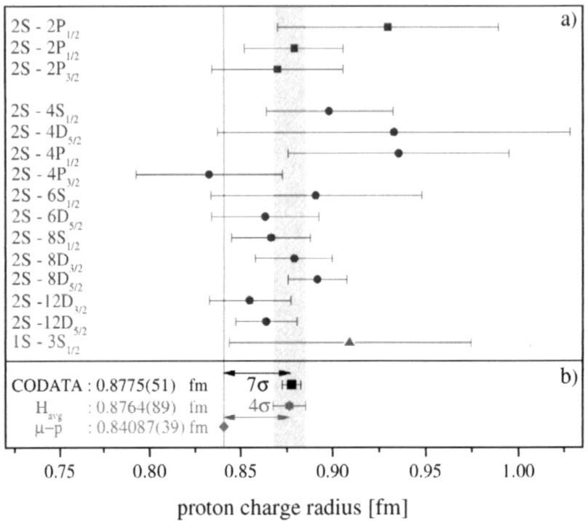

Figure 1. a) The proton charge radius obtained from precision spectroscopy of atomic hydrogen. Either radio frequency measurements of the 2S-2P Lamb shift (violet) or optical transition frequencies (blue from the 2S and green from the 1S state) are used. To extract the two parameters, the Rydberg constant and the proton charge radius, each of these measurements needs to be combined with another independent measurement, which is the 1S-2S transition frequency here. b) The analysis reveals a 4σ discrepancy between the hydrogen mean value (H_{avg}) and the value determined from laser spectroscopy of muonic hydrogen (μ–p). An even larger inconsistency of 7σ is obtained when including proton-electron scattering data. This CODATA analysis also uses deuterium data, that have only limited effect on the proton charge radius, because its nucleus is not a proton. A similar picture is obtained when plotting the corresponding results for the Rydberg constant.

3. Testing QED

While one would use the two hydrogen transition frequencies with the lowest uncertainty and the largest sensitivity to determine the parameters, testing the internal consistency of QED requires more input data. For this purpose one can use different pairs of hydrogen transition frequencies to determine values for r_p and R_∞ and verify that all pairs give the same result within their respective uncertainty. The analysis is done by combining the 1S-2S transition frequency, which has by far the lowest uncertainty, with each of the other measured transition frequencies that are listed in table XI of the CODATA 2010 report [4]. In this way one obtains 15 values for R_∞ and r_p of which only the latter is shown in fig. 1a. To check for hidden systematic effects in these data the weighted mean quadratic deviation from the weighted mean is used. The square root of this quantity is known as the Birge ratio R_B which has an expectation value of 1 for Gaussian distributed data and a variance of $2/(N-1)$ where $N = 15$ is the number of data points. For the proton charge radii determined from hydrogen one finds $R_\mathrm{B} = 0.70(38)$, so that the scattering of the data does not provide evidence for systematic uncertainties. However, systematic effects common to all measured transition frequencies can not be ruled out in this way.

Even though this simple approach uses the same value for the 1S-2S transition frequency several times for averaging, the influence on the overall result is insignificant because the 1S-2S contribution to the uncertainty is essentially zero. The proper method for obtaining best estimates for r_p and R_∞ using hydrogen spectroscopy data is a least square adjustment. The result of $r_p = 0.8764(89)$ fm with a Birge ratio of $R_\mathrm{B} = 0.96$ is given as "adjustment #8" in table XXXVIII of the CODATA 2010 report [4].

Another value for the proton charge radius is obtained from the measured 2S-2P$_{3/2}$ transition frequency in muonic hydrogen [2, 3]. Due to the enhanced sensitivity, moderate measurement accuracy and 5 of the 12 known digits of the Rydberg constant are sufficient to obtain the most precise value for the proton chare radius $r_p = 0.84087(39)$ fm. This value differs from the "hydrogenic" value by 4 combined standard deviations.

An even larger discrepancy of 7 combined standard deviations exists between the "muonic" value (see fig. 1b) and recommended the CODATA value that includes a selection of results from elastic electron proton scattering ("adjustment #3"). In the past electron proton scattering has produced inconsistent results for r_p. Depending on the evaluation method of the same data it may either support the "hydrogenic" [9] or the "muonic" [10] proton charge radius.

The discrepancy persists and QED is no longer consistent with experimental data, even when the electron proton scattering data are ignored. So the problem seems deeper and one may speculate about possible causes. Not only does it affect the value for r_p but likewise the value for R_∞. Nevertheless the problem has been referred to as the "proton size puzzle". From the experimental side one may ask by how many standard deviations certain measurements need to be off the true value to explain the discrepancy. It turns out that the measured 1S-2S transition frequency would have to be wrong by 4000 standard deviations while the measured 2S-2P transition frequencies in muonic hydrogen would have to be wrong by about 100 standard deviations. Unless a very large but yet unknown systematic effect is effective, it seems unlikely that an experimental problem there to be the cause the "proton size puzzle".

4. Improving the Hydrogen Input Data

The weakest experimental data are provided by the transitions in regular hydrogen that start from the metastable 2S level. This is because these transitions have larger line widths and/or are much more sensitive to stray electric fields. Still, as evident from fig. 1a, most individual values do not contradict the "muonic" value. Only when averaging the hydrogen data the discrepancy arises. For this reason we are seeking to reduce the uncertainty of at least one of those transition frequencies significantly. Depending on the outcome this may solve or reinforce the "proton size puzzle".

4.1. 2S-4P Transition

To generate a beam of metastable 2S atoms we are using two-photon laser excitation at 243 nm. Part of the experiment was used before to measure the 1S-2S transition frequency [1]. In all previous experiments that use the 2S as the ground state, electron impact excitation has been used to populate it. This method severely heats the atomic beam and even produces non-Maxwellian velocity distributions. It complicates a detailed line shape analysis which is essential to determine the line center much better than the line width. For the envisioned 2S-4P lines [11] we would like to find unperturbed line center within a few kHz in order to generate a value for the proton charge radius that can be decisive between the "muonic" and the "hydrogenic" value. This will require that we split the line better than parts in 10^3 of the natural line width of 12.9 MHz.

Another complication of the line shape arises from the unresolved hyperfine structure of the P states. Previous experiments have relied on selecting the hyperfine components by laser polarization. Thanks to the well resolved hyperfine splitting of the 1S-2S transition we can now selectively populate the 2S($F = 0$) hyperfine state. As the $F = 0 \rightarrow F = 0$ transition is strictly forbidden, independent of polarization, only one hyperfine component is left in the 2S-4P$_{1/2}$ manifold. Likewise, for the 2S-4P$_{3/2}$ transition only the $F = 0 \rightarrow F = 1$ can be excited. Avoiding the electron beam also helps to reduce the dc Stark effect by introducing less charges into the system.

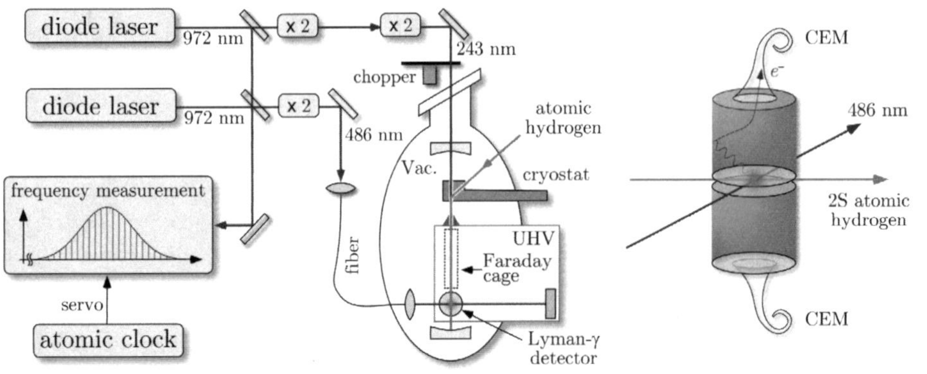

Figure 2. Left: An extended cavity laser diode at 972 nm [12], that is stabilized to a high finesse cavity (not shown) to obtain a 1 Hz line width [13], is frequency doubled twice (243 nm) to optically excite atomic hydrogen, that emerges from a 5.8 K nozzle, to the 2S state. A second similar laser is frequency doubled once (486 nm) and used to excite the 2S-4P transitions by illuminating the atomic beam at an angle as close as possible to 90°. An actively stabilized highly reflecting mirror ensures that the wavefronts of the beam coupled back into the fiber closely matches those of the forward wave. This suppresses the first-order Doppler effect to high degree. Right: Enlarged sketch of the Lyman-γ detector. Atoms in the 4P states quickly release a Lyman-γ photon at 97 nm. To obtain a large solid angle of detection, photons are not detected directly. Instead we detect photo-electron released from a cylindrical graphite coated surface. The photo-electrons are guided by electric fields to two channel electron multiplier (CEM) for detection. The atoms are shielded from the electric fields of the CEMs by a Faraday cage (not shown).

The apparatus that is sketched in fig. 2 uses two ultra stable diode lasers at 972 nm. One of them operates at a fixed frequency that is controlled by a frequency comb. Its 4-th harmonic runs collinearly with the atomic beam and drives the 1S-2S two-photon transition. From the previous 1S-2S experiments we know that the velocities of the resulting 2S atoms are essentially Maxwellian distributed with a temperature that is very close to 5.8 K which is the set temperature of the nozzle from which the 1S atoms emerge. The second laser at 972 nm is detuned by a about 2.3 GHz and scanned in frequency such that its second harmonic probes the 2S-4P transitions.

Transitions to the $4P_{1/2}$ and $4P_{3/2}$ levels are detected via their decay to the 1S ground state upon which they emit a Lyman-γ photon at 97 nm. These photons are detected via photo electrons that they generate inside two graphite coated cylinders. The photo electrons are guided to channel electron multipliers (CEM) with a suitable electric field configuration. With this detector we obtain a solid acceptance angle of almost 4π [14] while the large work function of graphite ($\approx$ 5 eV) suppresses the background from the 486 nm laser.

Almost all of the transitions from the 2S states that contribute to the "hydrogenic" proton charge radius are two-photon transitions to another S state or a D state. A dipole allowed one photon transition is subject to another set of possible systematic shifts. Therefore it is most suitable to investigate the "proton size puzzle". The largest among these systematic shifts is the first-order Doppler effect that is easily avoided in two-photon spectroscopy by using counter propagating laser beams. Similarly to the two-photon case we are using two counter propagating waves that cause opposite Doppler shifts which average to zero.

Taking the laser wavefront curvature into account the local Doppler shift is given by the projection of the velocity vector onto the normal of the local phase front. Therefore the back propagating wave has to match the forward wave in phase fronts and amplitude. To achieve this we couple the laser out of and back into the same single mode fiber with a waist at the flat high reflector whose tip and tilt are actively controlled (see fig. 2). At the same time we do not simply rely on perfectly matched counter propagating waves but measure the residual Doppler effect. To do that we employ time resolved detection: after turning off the 243 nm laser with a chopper we wait an adjustable time before opening the gate of the Lyman-γ detector. By this the fast atoms that could contribute to the signal have escaped and we are left with the slower ones within the Maxwell distribution. Observing the line center as a function of this delay we can set a very tight upper limit for the residual first-order Doppler effect.

We believe we have this effect under control and we are currently analysing other possible systematic shifts. Among them is the line pulling effect due to the quantum interference between different excitation and decay paths via the various Zeeman sublevels of the excited states [15]. As the emission pattern depends on these Zeeman sublevels, the observed line shift depends on the angle of the emitted photons relative to the polarization of the exciting laser. Recently this effect has been demonstrated experimentally with the D lines of Lithium [16]. This angle dependent line pulling effect exactly vanishes when collecting photons in all directions. With our Lyman-γ detector we hope to get close to this ideal situation.

4.2. 1S-3S/3D Transitions

Another transition that is under investigation in our lab for this purpose is the 1S-3S two-photon transition [17]. Its frequency has been measured by the group of F. Biraben with an uncertainty of 13 kHz [18]. Among the challenges here is the required wavelength of 205 nm that is difficult to produce as a continuous wave (cw). The only commercially available crystal that can generate it as a second harmonic is BBO. Not only does the required phase matching angle lie very close to 90° for that crystal, but it also suffers from photochemical reactions on the surfaces and from the photo-refractive effect. By cooling the crystal down to $-10°$ the phase matching angle shifts by about 1° and effective nonlinearity increases [19].

To deal with the remaining crystal issues we employ a mode-locked laser rather than a cw laser whose spectrum can be described as a frequency comb source. It features a high peak intensity and therefore greater nonlinear interactions and does not induce the photo-refractive effect. Thus, a deep ultraviolet frequency comb can, in general, be generated more efficiently than cw radiation. The comb modes can be expressed as $\omega_m = \omega_{ceo} + m\omega_r$, where ω_{ceo} is the carrier envelope offset frequency, ω_r the repetition rate of the laser and m an integer number. To drive a two-photon transition with a frequency comb the modes add pairwise to produce the transition energy $\hbar\omega_{eg}$. By tuning ω_{ceo} such that a comb mode ω_m corresponds to half the transition frequency, all mode pairs n which satisfy the relation $\omega_{eg} = \omega_{m-n}+\omega_{m+n}$ contribute to the excitation rate. This is sketched for $n = -1, 0, +1$ at the left side of fig. 3. The same applies if the two-photon transition occurs exactly between two comb modes thus the signal repeats with half the repetition rate when scanning ω_{ceo}. For Fourier-limited pulses, all frequency modes are properly phased and the excitation paths add up coherently. In this way a pulsed laser drives a two-photon transition as efficiently as a cw laser with the same average intensity while the spectral line width is limited by the width of a single comb mode rather than by the spectral envelope of the comb [20].

This two-photon direct frequency comb spectroscopy is not exactly free of the first-order Doppler effect. In the frequency domain one would argue that the contributing mode pairs do not have the same frequency in the lab frame. However, for each pair n there is a pair $-n$ such that the Doppler shift is balanced provided that the spectral envelopes of the counter propagating pulses are identical. This condition is closely matched by an enhancement cavity.

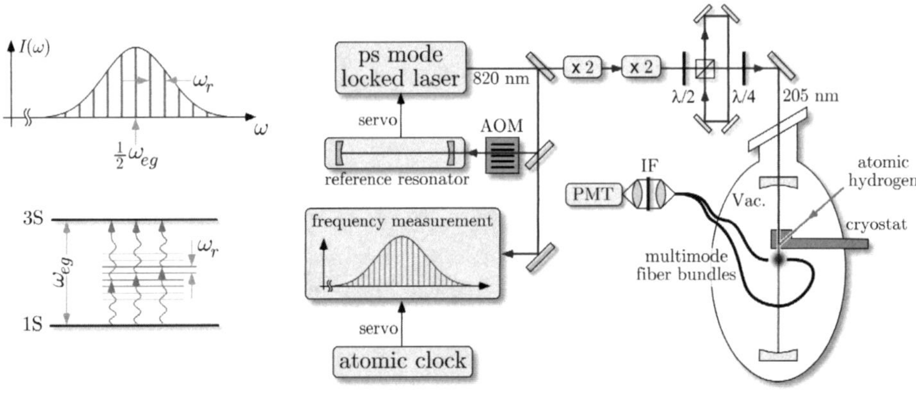

Figure 3. Left: spectral envelope of the frequency comb tuned to excite a two-photon transition at ω_{eg}. One of the modes is picked to measure the detuning, of which only values modulo the pulse repetition rate ω_r are known without additional information. On resonance, pairwise addition of properly phased modes provides an efficient excitation of the atoms. Three such mode pairs are shown. Right: A mode-locked Ti:sapphire laser provides Fourier-limited pulses of 1.3 ps pulse duration at $\omega_r = 2\pi \times 82$ MHz repetition frequency with an average output power of 1.8 W. This laser is stabilized to a reference cavity using the radio-frequency side band technique. A double-pass AOM serves for tuning while a self referenced frequency comb is used to measure the frequency of one of the modes of the ps laser. After two resonant frequency doubling stages 10 to 15 mW of power at 205 nm becomes available for spectroscopy. A third cavity inside a vacuum chamber enhances this power to 100 mW for spectroscopy. An interferometer allows to adjust the polarization and the position of the pulse collision point where the Doppler free excitation of the 1S-3S transition takes place. Bare multimode fibers are used to guide the emitted Balmer-α light through an interference filter (IF) onto a photomultiplier tube (PMT).

In the time domain one would argue that the finite excitation region leads to a time of flight broadening. In both cases the first-order Doppler effect gives rise to a line broadening rather than to a line shift. At a thermal velocity of about 350 m/s (5.8 K) a pulse duration of 1 ps gives rise to a time of flight (or Doppler) broadening of 500 kHz. This pulse duration is therefore well adapted to the natural line width of 1 MHz of the 1S-3S transition. A possible frequency chirp of the pulses will reduce the excitation rate but does not induce a line shift [21]. Furthermore, the ac-Stark shift is determined by the average power rather than by the peak power [22].

The experimental realization, that is sketched at the right side of fig. 3, uses a ps Ti:sapphire mode-locked laser at 820 nm that is frequency doubled twice to 205 nm using two enhancement cavities [19]. A Mach-Zehnder type interferometer generates double pulses and allows to adjust their pulse collision point inside the enhancement cavity where the Doppler free excitation takes place. With this interferometer the polarisation of the colliding pulses may be adjusted to enhance or suppress either the 1S-3S or the 1S-3D components. The signal is obtained by detecting the Balmer-α photons due to the decay of the 3S/3D to the 2P state at 656 nm. As shown in fig. 3 these photons are collected with multimode fiber bundles whose end facets are brought into close proximity of the pulse collision point. Using several of these fiber bundles we obtain a solid angle of about 20% of 4π.

Similar to the 2S-4P transitions we are currently investigating possible systematic frequency shifts. Besides the usual suspects we are also working out the effects of quantum interference.

4.3. Outlook

Two-photon direct frequency comb spectroscopy may be pushed to even shorter wavelengths. Pulsed lasers are readily up converted into the XUV spectral region and beyond by employing the process of high harmonic generation (HHG). With this it might become feasible to excite the 1S-2S transition in hydrogen like He^+ at 60.8 nm [23]. Besides proving another independent QED test this system has several advantages: As the power series in eqn.(1) is written for hydrogen, α has to be replaced by $Z\alpha$ for hydrogen like systems with nuclear charge $Z > 1$. Therefore He^+ provides a more sensitive test of the higher order terms in particular for yet uncalculated terms of order $(Z\alpha)^{6...7}$. On the experimental side He^+ holds the advantage that it can be held in an ion trap and sympathetically cooled by another ion with easy to generate cooling wavelength. This option is not yet available for hydrogen as the cooling wavelength of 121 nm is difficult to generate as a cw. The nuclear charge radius can be measured using muonic He^+. Preparations for such an investigation are under way at the Paul Scherer Institute, Switzerland [24]. A frequency comb at 60.8 nm has already been generated in our lab [25] albeit with power levels that are not yet sufficient to readily perform this spectroscopy.

References

[1] Ch. G. Parthey *et al.*, Phys. Rev. Lett. **107**, 203001 (2011).
[2] R. Pohl *et al.*, Nature **466**, 213 (2010).
[3] A. Antognini *et al.*, Science **339**, 417 (2013).
[4] P.J. Mohr, B.N. Taylor, and D.B. Newell, Rev. Mod. Phys. **84**, 1527 (2012).
[5] F. Biraben, Eur. Phys. J. SpecialTopics, **172**, 109 (2009).
[6] D. Hanneke, S. Fogwell, and G. Gabrielse, Phys. Rev. Lett. **100**, 120801 (2008).
[7] S. Rainville, J.K. Thompson, and D.E. Pritchard, Science **303** 334 (2004).
[8] D.L. Farnham, R.S. Van Dyck, and P.B. Schwinberg, Phys. Rev. Lett. **75**, 3598 (1995).
[9] J.C. Bernauer *et al.*, Phys. Rev. Lett. **105**, 242001 (2010).
[10] L.T. Lorenz, H.W. Hammer, and U.G. Meißner, Eur. Phys. J. A **48**:151 (2012).
[11] A. Beyer *et al.*, Ann. Phys. (Berlin), DOI 10.1002/andp.201300075
[12] N. Kolachevsky*et al.*, Opt. Lett. **36**, 4299 (2011).
[13] J. Alnis, A. Matveev, N. Kolachevsky, Th. Udem, and T.W. Hänsch, Phys. Rev A, **77**, 053809 (2008).
[14] A. Matveev *et al.*, Phys. Rev. Lett. **110**, 230801 (2013).
[15] M. Horbatsch and E.A. Hessels, Phys. Rev. A **82**, 052519 (2010).
[16] R.C. Brown *et al.*, Phys. Rev. A **87**, 032504 (2013).
[17] E. Peters, D.C. Yost, A. Matveev, T.W. Hänsch, and Thomas Udem, Ann. Phys. **525**, L29 (2013).
[18] O. Arnoult, F. Nez, L. Julien, and F. Biraben, Eur. Phys. J. D **60**, 243 (2010).
[19] E. Peters *et al.*, Opt. Express **17**, 9183 (2009).
[20] Y.V. Baklanov and V.P. Chebotayev, Appl. Phys. **12**, 97 (1977).
[21] S. Reinhardt, E. Peters, T.W. Hänsch, and Th. Udem, Phys. Rev. A **81**, 033427 (2010).
[22] P. Fendel, S.D. Bergeson, Th. Udem, and T.W. Hänsch, Opt. Lett. **32**, 701 (2007).
[23] M. Herrmann *et al.*, Phys. Rev. A **79**, 052505 (2009).
[24] A. Antognini *et al.*, Can. J. Phys. **89**, 47 (2011).
[25] A. Ozawa *et al.*, Phys. Rev. Lett. **100**, 253901 (2008).

Journal of Physics: Conference Series **467** (2013) 012004 doi:10.1088/1742-6596/467/1/012004

Ultrashort laser oscillator pulses meet nano-structures: from attosecond physics at metal tips to dielectric laser accelerators

Sebastian Thomas, Michael Krüger, John Breuer, Michael Förster, Peter Hommelhoff

Department of Physics, Friedrich Alexander University of Erlangen-Nuremberg, Staudtstr. 1, D-91058 Erlangen, and Max Planck Institute for Quantum Optics, Hans-Kopfermann-Str. 1, D-85748 Garching / Munich, EU

E-mail: peter.hommelhoff@fau.de

Abstract. We report on two experiments about the interaction of femtosecond laser pulses with nano-scale structures. In the first experiment, we have observed strong-field effects at nanometer sharp metal tips, such as rescattering, the high-energy cut-off and a strong carrier-envelope phase dependence. All these effects are hallmarks of attosecond physics. We have used the new understanding of these processes at a solid tip to build a near-field sensor with record sensitivity of 1 nm. In the second experiment, we have demonstrated the acceleration of non-relativistic electrons with the optical near-field at a photonic grating structure, with an acceleration gradient of 25 MeV/m. The grating serves to generate a near-field mode co-propagating with and efficiently imparting momentum onto the electron, which is impossible in vacuum over a macroscopic distance. For relativistic electrons we expect an acceleration gradient of more than 1 GeV/m with two-sided grating structures.

1. Introduction

Nanooptic or near-field optics opens possibilities that are impossible with common (far-field) optics. In the near-field modes with significant population can exist that are exponentially damped out in the far-field. We take advantage of nanooptics in two different settings. In the first part we will discuss our recent observations of strong-field effects driven by the near-field of and at a nano-scale metal tip. In the second part we will show how evanescent (near-field) modes can be utilized to accelerate electrons with the electro-magnetic light-field of femtosecond laser pulses. Whereas in the former case field enhancement at the tip allows us to work with low-power oscillator pulses and still reach the required peak intensities, so enables, in the second case, the grating the generation of the accelerating mode. This mode is co-propagating with the electrons and allows continuous and phase-synchronous momentum transfer from the laser beam to the electron beam.

2. Attosecond physics at metal nano-tips: the driven electron as near-field sensor

The recollision scenario lies at the core of strong-field and attosecond physics [1]. It describes that an electron, liberated by a strong laser pulse from an atom or molecule in the gas phase, can be driven back to the parent ion or molecule inside a single cycle of the laser field once the laser field has flipped sign. Several different processes can then happen: 1. The electron can recombine, upon which the energy that the electron has gained in the laser field is radiated away in the form of a high-harmonic photon. It is this process that coined the name of the field. 2. The electron can scatter elastically at the parent matter. In that case, the momentum vector just flips around, upon which the electron can gain more energy in the laser field. 3. Other processes, including inelastic scattering can take place.

In this contribution we focus on the second process. We observe and measure the electron, here emitted and re-colliding with a metal nano-tip. The metal tip here replaces the gas phase particle known from high harmonic generation physics.

Our basic observations have been described elsewhere [2, 3, 4, 5, 6], which is why we only briefly discuss the main features here. We have observed the tell-tale plateau in the electron spectra, which gives evidence of the elastic recollision [7]. Because the plateau indicates a process that is driven by the electric field of the laser pulse – as opposed to the pulse envelope – we have investigated the field dependence of the process. We have done this be recording spectra in a carrier-envelope phase (CEP) resolved way. These spectra reveal photon orders for certain settings of the CEP, whereas the photon orders are absent for other settings.

With the tree-step model (also known as simple man's model) [8] we are able to explain the spectra both qualitatively and quantitatively. The key point is that the photon picture is replaced by the notion of electronic matter wave interference in time and energy. Interference fringes in the energy domain show up if two emission windows in time exist during which the electron can be emitted. On the way to the detector, the two electronic matter wave packets spread out and overlap. This leads to interference fringes, which we detect with a retarding field spectrometer. In essence, in a CEP resolved measurement the photon picture can not explain the observed effects any longer: For certain settings of the CEP the emission process lacks any periodicity if only a single emission window exists. The intuitive process description therefore *requires* the matter wave picture and interference effects instead of photon orders. For CEP settings for which interference fringes exist, their spacing matches the photon picture because the emission windows in time are approximately spaced by one optical period $T \approx 2.8\,\mathrm{fs}$, which, according to Fourier, results in a fringe spacing of $\Delta E = h/T \approx 1.5\,\mathrm{eV}$.

The experimental data are well-supported by theory models [5, 6, 9, 10]. With the understanding of the dynamics, we can turn the perspective around and employ the re-colliding electron as a probe for the local electro-magnetic field in front of the tip [11]. This is based on the "10-U_p law" [12], which links the maximum classically allowable kinetic energy with the local intensity via

$$T_{\mathrm{cutoff}} = 10.007 U_p + 0.538\Phi.$$

Here, $U_p = e^2\lambda^2E^2/(16\pi^2mc^2)$ is the ponderomotive potential of the electron in the oscillating electric field with field strength E. e and m are the electron's charge and mass, and c is the speed of light. Φ is the workfunction of the tip material.

We measure T_{cutoff} and hence U_p and E. The field enhancement factor is the ratio between the measured peak field and the peak field in the bare laser focus, without the tip present. The results are shown for tungsten and gold tips as a function of tip radius in Fig. 1.

Also shown are FDTD simulation results for tungsten, gold and silver. For all materials, the field enhancement factor rises with decreasing tip radius, as can be expected from geometrical arguments. Interestingly, all materials exhibit a rather similar field enhancement factor, even though gold is considered to be a plasmonic material while tungsten is not. We note that we do observe a plasmon resonance in theory plots, however not for the dielectric function of gold for a wavelength of $\sim 800\,\mathrm{nm}$. Hence also for gold, the field enhancement effect is mainly geometric and thus broadband in nature. While these data show the efficacy of the method, more research is needed to figure out as to why larger field enhancement factors for gold are reported in the

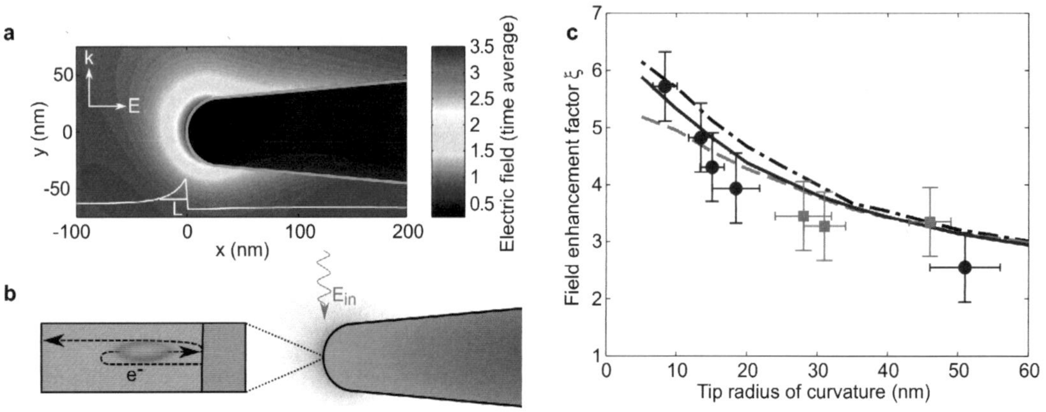

Figure 1. a) Field enhancement taking place at a nanoscale metal tip when laser pulses are focused at the tip. In the figure, the light is propagating in the y-direction with the polarization in the x-direction. The white line shows the decay of the field away from the tip on its symmetry axis. b) Conceptual picture of the electron dynamics in front of the tip. We utilize the electron wave packet to sample the field. The extension of the electron trajectory is less than 1 nm, hence the field can be approximated as constant. c) Results of the measurement of the electric field: the field enhancement factor is the ratio between the field obtained with the electron rescattering method and the field of the bare focus, without the tip present. Points are measurement data for tungsten (blue balls) and gold (red squares). The lines are theory curves obtained from extensive FDTD simulations. The solid blue line is for tungsten, the red dashed curve is for gold, and the dash-dotted is for silver.

literature [13]. We note that the spatial integration of 1 nm is almost one order of magnitude smaller than in the previously published record resolution work on for near-field sensing [14].

3. Dielectric laser acceleration with the inverse Smith-Purcell effect: efficient momentum transfer from a light field to electrons with the help of a transparent grating structure

In free space, momentum transfer from a oscillating field to a charged particle is not possible over extended time and length scales. This does not hold any more close to a boundary. With suitable grating structures, propagating electrons can be efficiently accelerated. This process becomes the more efficient, the closer the electrons' velocity approaches the speed of light.

Figure 2 shows the concept of the acceleration: The grating generates an evanescent mode that co-propagates with the electron and thus transfers momentum efficiently. Grating period, laser wavelength and electron velocity have to be matched, yielding the synchronicity condition

$$\lambda_p = m\beta\lambda.$$

Here, λ_p is the grating period, $m = 1, 2, 3\ldots$ and λ the laser wavelength. m represents the order of the spatial harmonics. In Figure 2, $m = 3$.

We send a beam of electrons, generated in and controlled with the help of a scanning electron microscope column, along a transparent grating structure. At the grating structure, the electron beam is overlapped with a laser beam. The laser beam is derived from a long-cavity Titanium:sapphire oscillator, with the following parameters: pulse duration of 110 fs, center

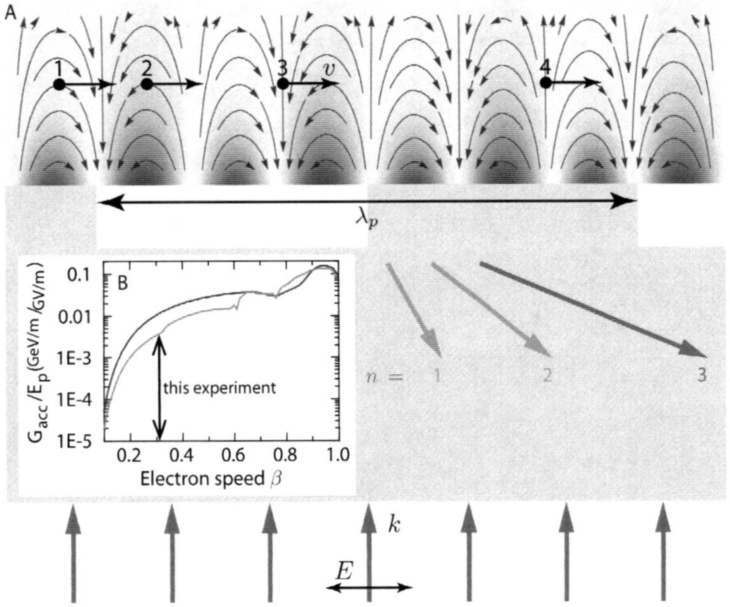

Figure 2. a) A plane wave propagates from top to bottom through a transparent grating structure (light blue). Electrons, indicated by the black dots, propagate parallel to the horizontal grating surface. If the grating period is matched to the speed of the electrons and the driving wavelengths, a mode co-propagating with the electrons at the surface at the grating is excited. It can efficiently transfer momentum to the electrons. In the figure, only electron 1 is accelerated. Electron 2 is decelerated, while electron 3 is deflected towards the grating and electron 4 away from the grating. b) The acceleration gradient $G_{\rm acc}$ for a given laser peak field $E_{\rm p}$ as function of the electron velocity.

wavelength of 787 nm, repetition rate of 2.745 MHz, average power of $\sim$ 350 mW. The focal parameters are: spot radius of 8.3 μm, peak electric field of 2.5 . . . 3 GV/m.

The grating is written into fused silica. The electron beam passes by with a distance down to $\sim$ 50 nm. Because of these small dimensions, we put the grating on top of a mesa structure such that the electron beam is only close to the glass where necessary.

We measure a maximum acceleration gradient of $\sim$ 25MeV/m. The polarization dependence, the dependence on incoming electron beam energy (synchronicity condition) and the dependence on the distance of the electron beam from the grating surface all agree with what is expected from this near-field based scheme of electron acceleration. We are currently in the process of writing up the results in greater detail and expect them to be submitted soon.

In summary, near-field based experiments with femtosecond oscillator pulses at nanoscale tips and at transparent grating structures have allowed the observation of novel effects, such as strong-field physics at solids and electron acceleration right with the electro-magnetic field of laser pulses. The latter work represents the first observation of the inverse Smith-Purcell effect in the optical regime.

References

[1] P. B. Corkum and F. Krausz, Nat. Phys. **3**, 381 (2007).
[2] M. Schenk, M. Krüger, and P. Hommelhoff, Phys. Rev. Lett. **105**, 257601 (2010).
[3] M. Krüger, M. Schenk, and P. Hommelhoff, Nature **475**, 78 (2011).

[4] M. Krüger, M. Schenk, M. Förster, and P. Hommelhoff, J. Phys. B **45**, 074006 (2012).
[5] M. Krüger, M. Schenk, P. Hommelhoff, G. Wachter, C. Lemell, and J. Burgdörfer, New J. Phys. **14**, 085019 (2012).
[6] G. Wachter, C. Lemell, J. Burgdörfer, M. Schenk, M. Krüger, and P. Hommelhoff, Phys. Rev. B **86**, 035402 (2012).
[7] G. G. Paulus, W. Becker, W. Nicklich, and H. Walther, J. Phys. B: At. Mol. Opt. Phys. **27**, L703 (1994).
[8] P. B. Corkum, Phys. Rev. Lett. **71**, 1994 (1993).
[9] S. V. Yalunin, M. Gulde, and C. Ropers, Phys. Rev. B **84**, 195426 (2011).
[10] S. V. Yalunin, G. Herink, D. R. Solli, M. Krüger, P. Hommelhoff, M. Diehn, A. Munk, and C. Ropers, Ann. Phys. **525**, L12 (2013).
[11] S. Thomas, M. Krüger, M. Förster, M. Schenk, and P. Hommelhoff (submitted); preprint at http://arxiv.org/abs/1209.5195 (2013).
[12] M. Busuladžić, A. Gazibegović-Busuladžić, and D. B. Milošević, Las. Phys. **16**, 289 (2006).
[13] A. Hartschuh, Angewandte Chemie International Edition **47**, 8178 (2008).
[14] M. B. Raschke, L. Molina, T. Elsaesser, D. H. Kim, W. Knoll, and K. Hinrichs, ChemPhysChem **6**, 2197 (2005).

Mixed-field orientation of a thermal ensemble of linear polar molecules

Juan J. Omiste and Rosario González-Férez

Instituto Carlos I de Física Teórica y Computacional and Departamento de Física Atómica, Molecular y Nuclear, Universidad de Granada, 18071 Granada, Spain

E-mail: rogonzal@ugr.es

Abstract. We present a theoretical study of the impact of an electrostatic field combined with nonresonant linearly polarized laser pulses on the rotational dynamics of a thermal ensemble of linear molecules. We solve the time-dependent Schrödinger equation within the rigid rotor approximation for several rotational states. Using the carbonyl sulfide (OCS) molecule as a prototype, the mixed-field orientation of a thermal sample is analyzed in detail for experimentally accessible static field strengths and laser pulses. We demonstrate that for a characteristic field configuration used in current mixed-field orientation experiments, a significant orientation is obtained for rotational temperatures below 0.7 K as well as using stronger dc fields.

1. Introduction

The availability of samples of oriented molecules plays a crucial role in the control of chemical reaction dynamics [1, 2, 3, 4, 5], in molecular frame investigations such as photoelectron angular distributions [6, 7, 8], in high-order harmonic generation [9, 10, 11], or in the measurements of molecular properties such as the permanent dipole moment [12]. The mixed-field orientation of polar molecules via the interaction with an electric field and a nonresonant laser field is a widespread technique to produce such samples of oriented molecules. This method was proposed by Friedrich and Herschbach [13, 14], and is based on the dc-field induced coupling between the nearly degenerate pair of states with opposite parity forming the tunneling doublets in the strong laser field regime. A recent experimental and theoretical study has proven that under ns laser pulses the weak dc field orientation is not, in general, adiabatic, and that a time-dependent description of the mixed-field orientation process is required to explain the experimental results [15, 16]. Thus, depending on the field configuration, the orientation of a rotational state could be significantly smaller than the adiabatic prediction. In addition, not all the states present a right-way orientation, and some of them are antioriented.

In a thermal ensemble of molecules, the combination of these right- and wrong-way oriented states gives rise to a weakly oriented molecular beam [17, 18]. An enhancement of the orientation could be achieved by employing either lower rotational temperatures or quantum-state selected molecular beams. By using inhomogeneous electric fields, the amount of populated states is significantly reduced creating a quantum-state selected molecular beam, and achieving with this beam an unprecedented degree of orientation [19, 20, 21]. Cold molecular beams, with typical temperatures of the order of 1 K, are created in supersonic expansions of molecules seeded in an inert atomic carrier gas [22]. Depending on the rotational constant, the molecules could still be distributed over a large number of rotational states in these thermal ensembles. In the present work, we investigate the mixed-field orientation of a thermal sample of polar molecules as the rotational temperature is varied. Our aim is to find the temperature at which the thermal ensemble shows a similar orientation as the quantum-state selected molecular beam.

Herein, we consider a polar linear molecule exposed to an electric field combined with a nonresonant laser pulse, and provide a detailed theoretical analysis of the mixed-field orientation of a thermal sample of this molecule. To do so, we solve the time-dependent Schrödinger equation within the rigid rotor approximation for a large set of rotational states. Taking as prototype example the OCS molecule, we explore the mixed-field orientation as a function of the rotational temperature of the thermal sample for several experimental field configurations. We show that to achieve a significant orientation, rotational temperatures around 0.6 K and 1 K are required if either a weak or strong dc fields are applied, respectively. We also present the orientation of individual states and, for some of them, analyze the projections of the time-dependent wave functions on the corresponding adiabatic basis.

The paper is organized as follows: In Sec. 2 we describe the Hamiltonian of the system and the orientation of a molecular thermal ensemble. The mixed-field orientation of the thermal ensemble as a function of the rotational temperature is analyzed in Sec. 3. The conclusions are given in Sec. 4.

2. The Hamiltonian and the orientation of a thermal ensemble

We consider a polar linear molecule exposed to a homogeneous static electric field and a nonresonant linearly polarized laser pulse. In the framework of the rigid rotor approximation, the Hamiltonian of this system reads

$$H(t) = H_r + H_{\rm s}(t) + H_{\rm L}(t), \tag{1}$$

where H_r is the field-free Hamiltonian

$$H_r = B\mathbf{J}^2, \tag{2}$$

with $\mathbf{J}$ being the total angular momentum operator and B the rotational constant. The interactions with the electric and laser fields are $H_{\rm s}(t)$ and $H_{\rm L}(t)$, respectively.

The dc field $\mathbf{E}_{\rm s}(t)$ forms an angle β with the Z-axis and is contained in the XZ-plane of the laboratory fixed frame (LFF) (X, Y, Z). The dipole coupling with this field reads as

$$H_{\rm s}(t) = -\boldsymbol{\mu}\cdot\mathbf{E}_{\rm s}(t) = -\mu \mathrm{E}_{\rm s}(t)\cos\theta_{\rm s} \tag{3}$$

with $\mathbf{E}_{\rm s}(t) = \mathrm{E}_{\rm s}(t)(\sin\beta\hat{X} + \cos\beta\hat{Z})$, and $\mathrm{E}_{\rm s}(t)$ being the electric field strength. The angle between the dipole moment $\boldsymbol{\mu}$ and $\mathbf{E}_{\rm s}(t)$ is $\theta_{\rm s}$, and $\cos\theta_{\rm s} = \cos\beta\cos\theta + \sin\beta\sin\theta\cos\phi$. The angles $\Omega = (\theta, \phi)$ are the Euler angles, which relate the laboratory and molecular fixed frames. The molecule fixed frame (MFF) (X_M, Y_M, Z_M) is defined so that the molecular permanent dipole moment $\boldsymbol{\mu}$ is parallel to the Z_M-axis. Based on the mixed-field orientation experiments [20, 21, 15], the dc field is switched on first increasing its strength linearly with time. We ensure that this turning-on process is adiabatic, and once the maximum strength $\mathrm{E}_{\rm s}$ is achieved, it is kept constant.

The polarization of the nonresonant laser field is taken parallel to the Z-axis. Thus, the interaction of the nonresonant laser field with the molecule is written as [23]

$$H_{\rm L}(t) = -\frac{\mathrm{I}(t)}{2c\epsilon_0}\Delta\alpha\cos^2\theta, \tag{4}$$

Journal of Physics: Conference Series **467** (2013) 012005 doi:10.1088/1742-6596/467/1/012005

where $\Delta\alpha$ is the polarizability anisotropy, $\mathrm{I}(t)$ is the intensity of the laser, c is the speed of light and ϵ_0 is the dielectric constant. Note that in Eq. (4) the term $-\alpha_\perp \mathrm{I}(t)/2c\epsilon_0$ has been neglected because it represents only a shift in the energy. The laser is a Gaussian pulse with intensity $\mathrm{I}(t) = \mathrm{I}_0 \exp\left(-t^2/2\sigma^2\right)$, I_0 is the peak intensity, and σ is related with the full width half maximum (FWHM) $\tau = 2\sqrt{2\ln 2}\sigma$. When the nonresonant laser field is turned on the interaction due to this field is much weaker than the coupling with the dc field.

The time-dependent Schrödinger equation associated to the Hamiltonian (1) is solved by means of a second-order split-operator technique [24], combined with the discrete-variable and finite-basis representation methods for the angular coordinates [25, 26, 27, 28]. The basis is formed by the spherical harmonics $Y_{JM}(\Omega)$, which are the eigenstates of the field-free Hamiltonian (2). J and M are the rotational and magnetic quantum numbers, respectively. At time t, the time-dependent states will be labelled as $|J, M, l\rangle^{\beta}{}_t$ with $l = e$ and o indicating even or odd parity with respect to the XZ-plane, respectively. The labels J, M and l refer to the field-free quantum numbers to which they are adiabatically connected and they depend on the way the fields are turned on [29].

We consider a thermal sample of molecules and investigate its mixed-field orientation at $t = 0$ once the peak intensity I_0 has been achieved. For a rotational temperature T, the orientation of a thermal distribution is given by

$$\langle\langle\cos\theta\rangle\rangle_T = \sum_{J=0}^{\infty}\sum_{M=-J}^{J} W_J^T \langle\cos\theta\rangle_{JM}$$

where the orientation of the field-dressed state $|J, M, l\rangle^{\beta}{}_0$ is $\langle\cos\theta\rangle_{JMl} = {}_0{}^{\beta}\langle J, M, l|\cos\theta|J, M, l\rangle^{\beta}{}_0$. The thermal weight of the field-free state $|J, M, l\rangle$ is

$$W_J^T = \frac{e^{\frac{-J(J+1)B}{k_B T}}}{W^T} \qquad W^T = \sum_{J=0}^{\infty}(2J+1)e^{\frac{-J(J+1)B}{k_B T}} \tag{5}$$

with k_B being the Boltzman constant.

In many mixed-field orientation experiments, the degree of orientation is measured by the ion imaging method [20, 21]. The up/down symmetry of the 2D-images of the ionic fragments is experimentally quantified by the ratio $\mathrm{N_{up}}/\mathrm{N_{tot}}$, with $\mathrm{N_{up}}$ being the amount of ions in the upper part of the screen plane, and $\mathrm{N_{tot}}$ the total number of detected ions. In order to compare with the experimental results [15], we also compute the orientation ratio $\mathrm{N_{up}}/\mathrm{N_{tot}}$, of this thermal sample on a 2D screen perpendicular to the electric field axis. This is defined as

$$\frac{\mathrm{N_{up}}}{\mathrm{N_{tot}}} = \sum_{J}\sum_{M=-J}^{J} W_J^T \frac{\mathrm{N_{up}^{JM}}}{\mathrm{N_{tot}^{JM}}}$$

where

$$\mathrm{N_{up}^{JM}} = \int_{y^2+z^2\leq 1}\int_{z\geq 0} P_{JM}(y,z)\, dydz,$$

and

$$\mathrm{N_{tot}^{JM}} = \int_{y^2+z^2\leq 1} P_{JM}(y,z)\, dydz$$

with $P_{JM}(y,z)$ being the projection on a 2D screen perpendicular to the electric field axis of the probability density associated to the state $|J, M, l\rangle^{\beta}{}_0$ [30], which includes the alignment selectivity of the probe laser. y and z are the abscissa and ordinate of a 2D coordinate system

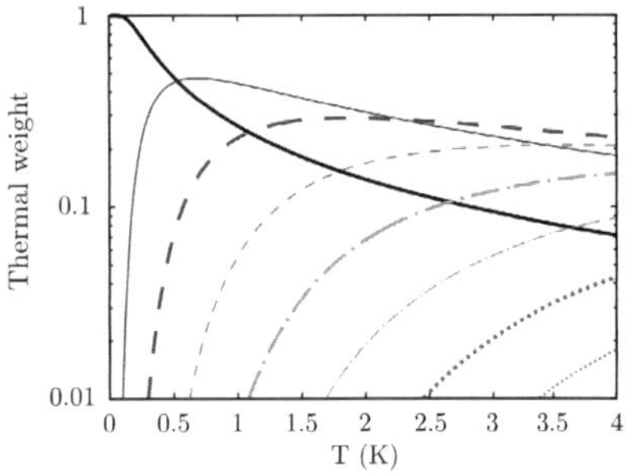

Figure 1. For OCS, thermal weights as a function of the temperature for several J-manifolds: $J = 0$ (thick solid line), $J = 1$ (thin solid line), $J = 2$ (thick dashed line), $J = 3$ (thin dashed line), $J = 4$ (thick dot-dashed line), $J = 5$ (thin dot-dashed line), $J = 6$ (thick dotted line) and $J = 7$ (thin dotted line).

centered on the screen, due to their relation with the Euler angles (θ, ϕ) their values are restricted to $y^2 + z^2 \leq 1$ [30].

To rationalize the mixed-field orientation results and illustrate the adiabaticity of this process, the time-dependent wave function is projected on the field-dressed adiabatic states

$$|J, M, l\rangle_t^\beta = \sum_{j=0}^{N} \sum_{m_j=-j}^{j} C_{jm_jl'}(t)\,|j, m_j, l'\rangle_{\mathrm{p}}^\beta \tag{6}$$

with $C_{jm_jl'}(t) = {}^\beta{}_{\mathrm{p}}\langle j, m_j, l'|\,JMl\rangle_t^\beta$. This adiabatic basis is formed by the eigenstates $|j, m_j, l\rangle_{\mathrm{p}}^\beta$ of the adiabatic Hamiltonian, i. e., the Hamiltonian (1) with constant electrostatic field $\mathrm{E_s}$ and constant laser intensity $\mathrm{I} = \mathrm{I}(t)$. For each time t, the time-independent Schrödinger equation is solved by expanding the wave function in a basis formed by linear combinations of spherical harmonics that respects the symmetries of the system. Note that for $|J, M, l\rangle^\beta{}_0$, the closer $|C_{JMl}|^2$ to one the more adiabatic is the mixed-field orientation process.

3. Results

In this work, we use the OCS molecule as prototype. The rotational constant of OCS is $B = 0.20286$ cm^{-1}, the permanent dipole moment $\mu = 0.71$ D and the polarizability anisotropy $\Delta\alpha = 4.04$ Å^3. In Fig. 1, we present the thermal weights of several rotational manifolds $(2J + 1)W_J$, see Eq. (5). Due to the large rotational constant of OCS, the field-free energy splittings are large, and then, the thermal samples with $T \lesssim 1$ K are dominated by the $J = 0$ and $J = 1$ manifolds. Indeed, the relative weights of the states with $J = 0$ and $J = 1$ are $W_0 = 47.8\%$ and $W_1 = 44.7\%$ at $T = 0.5$ K, and $W_0 = 99.1\%$ and $W_1 = 0.9\%$ at $T = 0.1$ K. In our calculations, the thermal sample includes rotational states with $J \leq 9$, and we have ensured that the contribution of higher excitations can be neglected.

We first consider the OCS molecules exposed to an electric field and linearly polarized laser pulse, with both fields parallel to the LFF Z-axis. For several field configurations, we present in Fig. 2 the orientation cosine of the thermal ensemble as a function of the temperature for $\mathrm{E_s} = 300\,\mathrm{V/cm}$. Note the different scales used in each panel.

For this weak dc field, a significant orientation is only achieved if the rotational temperature is below 0.5 K, and the Gaussian pulse has $\tau = 10$ ns, e. g., for the peak intensities $\mathrm{I_0} = 10^{12}\,\mathrm{W/cm^2}$ and $5 \times 10^{11}\,\mathrm{W/cm^2}$ we obtain $\langle\langle\cos\theta\rangle\rangle_T \gtrsim 0.5$. Using 1 ns Gaussian pulse, the orientation of the thermal sample is very small because the rotational states are weakly oriented, for instance, they satisfy $|\langle\cos\theta\rangle_{JMl}| < 0.13$ for $\mathrm{I_0} = 10^{12}\,\mathrm{W/cm^2}$ and $\mathrm{I_0} = 5 \times 10^{11}\,\mathrm{W/cm^2}$; whereas for $\mathrm{I_0} = 10^{11}\,\mathrm{W/cm^2}$, we obtain $\langle\cos\theta\rangle_{00e} = 0.24$ for the ground state. For these three FWHM, we encounter that a pulse with peak intensity $\mathrm{I_0} = 5 \times 10^{11}\,\mathrm{W/cm^2}$ gives rise to a larger orientation than one with $\mathrm{I_0} = 10^{12}\,\mathrm{W/cm^2}$, this is counterintuitive to what is expected in the adiabatic

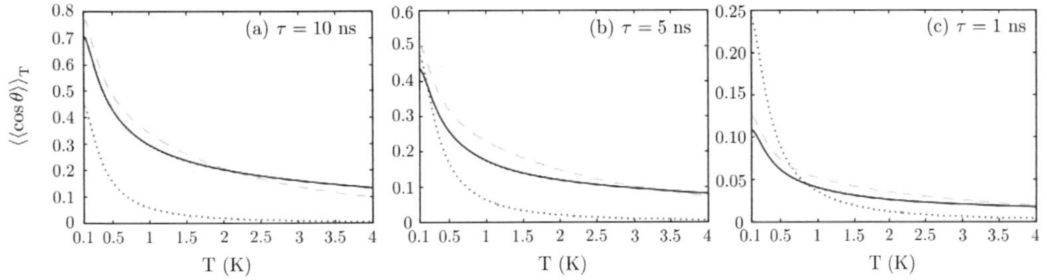

Figure 2. Orientation of a OCS thermal sample $\langle\langle\cos\theta\rangle\rangle_T$ as a function of the temperature for Gaussian pulses with $\tau = 10$ ns, $\tau = 5$ ns, and $\tau = 1$ ns and peak intensities $I_0 = 10^{12}\,\mathrm{W/cm^2}$ (thick solid line), $I_0 = 5\times10^{11}\,\mathrm{W/cm^2}$ (dashed line) and $I_0 = 10^{11}\,\mathrm{W/cm^2}$ (dotted line). The field configuration is $E_s = 300\,\mathrm{V/cm}$ and $\beta = 0°$.

limit. This phenomenon can be explain by the non-adiabaticity of the mixed-field orientation process [15, 16], and can be rationalized in terms of the orientation of the individual levels. In Fig. 3, we present the orientation cosine of the field-dressed states $|J,|M|,e\rangle^0{}_0$ at $t = 0$ for two 10 ns Gaussian pulses with $I_0 = 10^{12}\,\mathrm{W/cm^2}$ and $I_0 = 5\times10^{11}\,\mathrm{W/cm^2}$. In these plots, we observe that the levels $|J,M,e\rangle^0{}_0$ and $|J+1,M,e\rangle^0{}_0$, which form a pendular doublet, are oriented and antioriented, respectively. The $5\times10^{11}\,\mathrm{W/cm^2}$ pulse is not strong enough to affect the rotational dynamics in the excited rotational states with $J \geq 5$. The pulse with the strongest intensity $I_0 = 10^{12}\,\mathrm{W/cm^2}$ provokes a large orientation on highly excited states with $J \leq 7$. However, for the levels with $J \leq 3$, i. e., those that are important on the cold regime, the $5\times10^{11}\,\mathrm{W/cm^2}$ pulse gives rise to a larger orientation compared to the $10^{12}\,\mathrm{W/cm^2}$ one. In the parallel field configuration, the population transfer between the two levels forming the doublets in the pendular regime is the only source of nonadiabatic effects in the field-dressed dynamics [15, 16]. For these levels, the population transfer to the neighboring state as the pendular pair is formed is the largest for the strongest laser. For the ground state, at $t = 0$ we obtain that the population of the adiabatic state $|0,0,e\rangle^0_{\mathrm{p}}$ is $|C_{00e}|^2 = 0.87$ and 0.91 with $I_0 = 10^{12}\,\mathrm{W/cm^2}$ and $I_0 = 5\times10^{11}\,\mathrm{W/cm^2}$, respectively. As a consequence, the orientation is smallest for $I_0 = 10^{12}\,\mathrm{W/cm^2}$, and, therefore, the thermal ensemble is less oriented. By increasing the temperature, the contribution of excited rotational states becomes important, and the thermal ensemble in a $I_0 = 10^{12}\,\mathrm{W/cm^2}$ pulse shows the largest orientation.

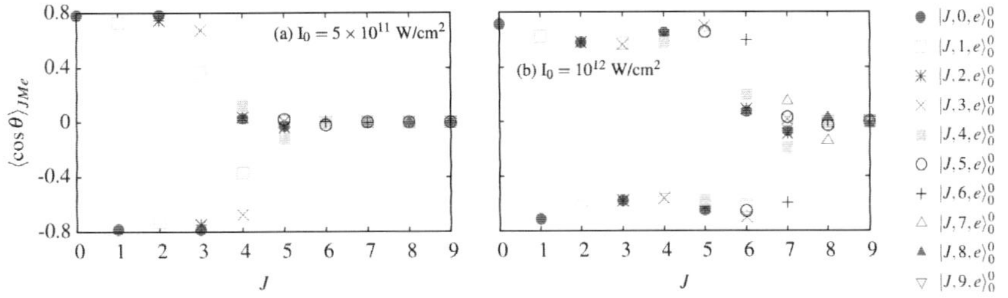

Figure 3. Orientation cosines $\langle\cos\theta\rangle_{JMl}$ at $t = 0$ of the states $|J,|M|,e\rangle^0{}_0$ versus the field-free rotational quantum number J. The Gaussian pulses have $\tau = 10$ ns, and peak intensities (a) $I_0 = 5\times10^{11}\,\mathrm{W/cm^2}$ and (b) $I_0 = 10^{12}\,\mathrm{W/cm^2}$. The field configuration is $E_s = 300\,\mathrm{V/cm}$ and $\beta = 0°$.

Now, we consider that the electric field is tilted an angle $\beta = 30°$ with respect to the

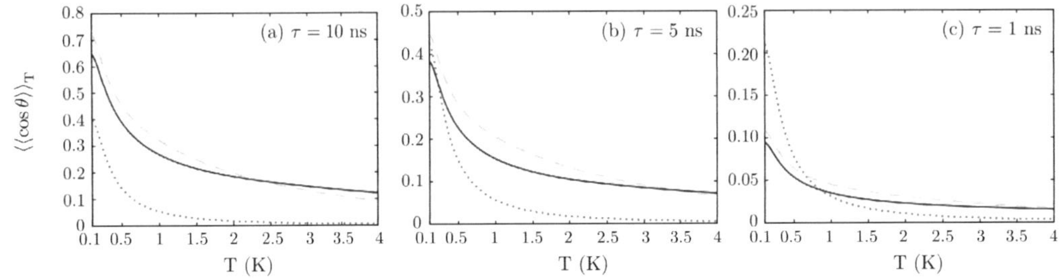

Figure 4. Same as Fig. 2 but for $\beta = 30°$.

polarization axis of the laser pulse, that is the LFF Z-axis. For several field configurations, we present in Fig. 4 the orientation cosine of the thermal ensemble as a function of the temperature for $E_s = 300\,\mathrm{V/cm}$. Compared to the parallel field case, the orientation is reduced. For tilted fields, there are two main sources of nonadiabatic effects in the field-dressed dynamics: i) the transfer of population taking place when the quasidegenerate pendular doublets are formed as the laser intensity is increased; ii) at weak laser intensities, there is also population transfer due to the splitting of the states within a J-manifold now having the same symmetry. In addition, avoided crossings might be encountered as $I(t)$ is enhanced. The diabatic or adiabatic character of these avoided crossings depends on the field configuration and on the state. Hence, for a certain field configuration, the orientation of the individual states is smaller for $\beta = 30°$ than for $\beta = 0°$. This reduction of the orientation is illustrated for the rotational states $|J, M, e\rangle^0{}_0$ in Fig. 5 for two 10 ns Gaussian pulses. For $I_0 = 5 \times 10^{11}\,\mathrm{W/cm^2}$, only the states $|0, 0, e\rangle^0{}_0$ and $|3, 1, e\rangle^0{}_0$ present a strong orientation with $|\langle\cos\theta\rangle_{JMl}| > 0.6$, whereas for $I_0 = 10^{12}\,\mathrm{W/cm^2}$ only the ground state is strongly oriented. The other levels present a moderate or even small orientation. Due to the population redistribution within a J-manifold at weak intensities, the two levels forming a pendular doublet do not possess the same orientation $|\langle\cos\theta\rangle_{JMl}|$ but in opposite directions as occurs in the parallel field configuration.

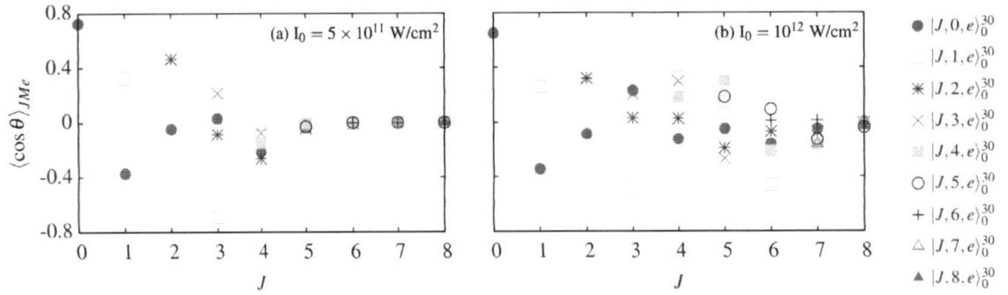

Figure 5. For the states $|J, |M|, e\rangle^{30}{}_0$, orientation cosines $\langle\cos\theta\rangle_{JMl}$ at $t = 0$ versus the field-free rotational quantum number J. The Gaussian pulses have $\tau = 10$ ns and peak intensities (a) $I_0 = 5 \times 10^{11}\,\mathrm{W/cm^2}$ and (b) $I_0 = 10^{12}\,\mathrm{W/cm^2}$. The field configuration is $E_s = 300\,\mathrm{V/cm}$ and $\beta = 30°$.

For the state $|2, 0, e\rangle^{30}{}_t$, we illustrate its rotational dynamics by presenting the projections of the time-dependent wave function in terms of the adiabatic states in Fig. 6(a) for a 10 ns pulse with $I_0 = 10^{12}\,\mathrm{W/cm^2}$, $E_s = 300\,\mathrm{V/cm}$ and $\beta = 30°$. The switching on of the electric field has been adiabatic and the level $|2, 0, e\rangle^{30}_{\mathrm{p}}$ is the only one populated when the laser pulse is turned on. At weak laser intensities, the three states with the same symmetry in the $J = 2$ manifold, that is $|2, 0, e\rangle^{30}_{\mathrm{p}}$, $|2, 1, e\rangle^{30}_{\mathrm{p}}$ and $|2, 2, e\rangle^{30}_{\mathrm{p}}$, are driven apart: $|C_{20e}(t)|^2$

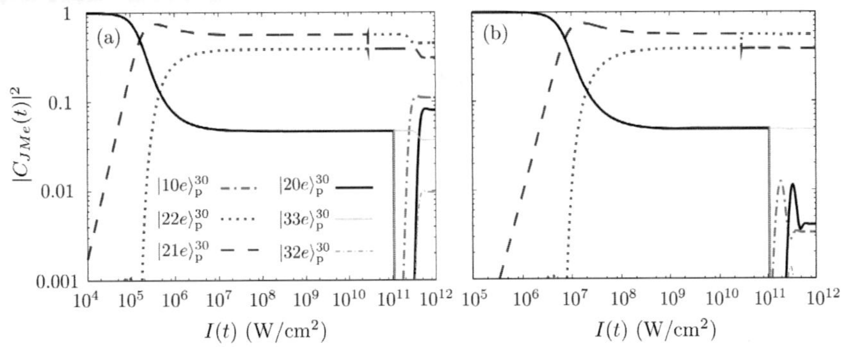

Figure 6. For the state $|2,0,e\rangle^{30}{}_t$, we present the squares of the projections of the time dependent wave function onto the adiabatic pendular states versus the laser intensity I(t), for dc field strengths (a) $E_s = 300$ V/cm and (b) $E_s = 2$ kV/cm. The Gaussian pulse has $\tau = 10$ ns and peak intensity $I_0 = 10^{12}$ W/cm^2, and the fields are tilted an angle $\beta = 30°$.

decreases as I(t) is increased, whereas $|C_{21e}(t)|^2$ and $|C_{22e}(t)|^2$ increase. For a wide range of laser intensities, these three coefficients keep their values constant. Around $I(t) \approx 2.84 \times 10^{10}$ W/cm^2, the states $|2,1,e\rangle_p^{30}$ and $|2,2,e\rangle_p^{30}$ suffer an avoided crossings, which is crossed diabatically and the population of these two adiabatic levels is interchanged. Another diabatic avoided crossing is encountered around $I(t) \approx 1.09 \times 10^{11}$ W/cm^2, and the involved states $|2,0,e\rangle_p^{30}$ and $|3,3,e\rangle_p^{30}$ interchange their population. Upon further increasing I(t), the pendular doublets start to form, the coupling between the two involved states increases, and there is a new population redistribution. In this figure, it is appreciated how the different pendular doublets are formed sequentially according to their energy. The first one involves the states $|1,0,e\rangle_p^{30}$ and $|2,2,e\rangle_p^{30}$, the next one $|2,1,e\rangle_p^{30}$ and $|2,0,e\rangle_p^{30}$, and the third one in this figure $|3,3,e\rangle_p^{30}$ and $|3,2,e\rangle_p^{30}$. At $t = 0$, the contribution of the adiabatic states to the field-dressed wave function is $|C_{22e}(0)|^2 = 0.11$, $|C_{22e}(0)|^2 = 0.45$, $|C_{21e}(0)|^2 = 0.31$, $|C_{20e}(0)|^2 = 0.08$, $|C_{33e}(0)|^2 = 0.04$ and $|C_{32e}(0)|^2 = 0.01$. As a consequence of this population redistribution, at $t = 0$ the state $|2,0,e\rangle^{30}{}_0$ is weakly antioriented $\langle\cos\theta\rangle_{20e} = -0.089$, whereas in the adiabatic prediction present a strong anti-orientation $\langle\cos\theta\rangle_{20e} = -0.886$. Analogously, other features of the system such as the energy, alignment, and hybridization of the angular motion are also affected by this population redistribution and do not resemble the adiabatic results.

For $\beta = 30°$, the orientation ratio N_{up}/N_{tot} is presented in Fig. 7. To compute N_{up}/N_{tot} we have used a probe laser linearly polarized along the vertical axis of the screen detector as in the experiments [15]. In these results, we have neglected the volume effect [30], we should mention that by including it the value of N_{up}/N_{tot} will be reduced.

In recent experiments [15], for a state selected molecular beam of OCS, 92% in $|0,0,e\rangle^0{}_0$, 4% in $|1,1,e\rangle^0{}_0$ and 4% in $|1,1,o\rangle^0{}_0$, an orientation ratio of $N_{up}/N_{tot} = 0.73$ was achieved using a 8 ns YAG laser with $I_0 = 9.1 \times 10^{11}$ W/cm^2, $E_s = 286$ V/cm and $\beta = 30°$. Using a 10 ns pulse, similar results for the orientation ratio of the thermal ensemble are reached if the rotational temperature is sufficiently low. For instance, $N_{up}/N_{tot} \gtrsim 0.73$ for $T \lesssim 0.65$ K and 0.46 K with peak intensities $I_0 = 5 \times 10^{11}$ W/cm^2 and $I_0 = 10^{12}$ W/cm^2, respectively. At $T = 0.65$ K, the field-free thermal ensemble is formed by 38.56% OCS in its ground state, 47.19% in $J = 1$ and 13% in $J = 2$; whereas for $T = 0.46$ K, 51.08% have $J = 0$, 43.07% $J = 1$, and 5.7% $J = 2$. For $\tau = 5$ ns, only when more than 95% of OCS molecules are in the ground state and $I_0 = 5 \times 10^{11}$ W/cm^2 we obtain a similar orientation ratio as in the experiment. By reducing

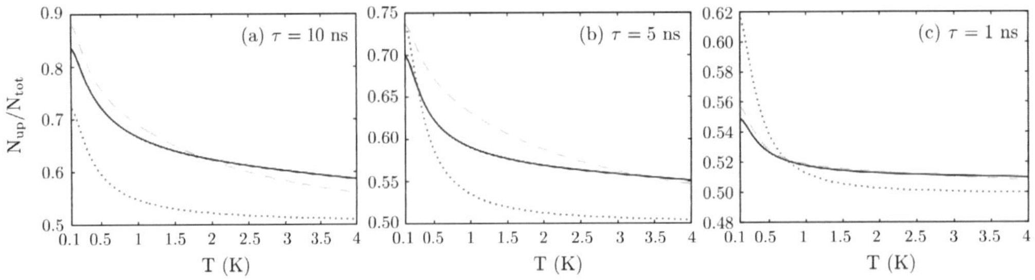

Figure 7. For the OCS thermal sample, we present the orientation ratio N_{up}/N_{tot} as a function of the temperature for Gaussian pulses with FWHM $\tau = 10$ ns, $\tau = 5$ ns, and $\tau = 1$ ns and peak intensities $I_0 = 10^{12}\,W/cm^2$ (thick solid line), $I_0 = 5 \times 10^{11}\,W/cm^2$ (dashed line) and $I_0 = 10^{11}\,W/cm^2$ (dotted line). The field configuration is $E_s = 300\,V/cm$ and $\beta = 30°$.

the FWHM to 1 ns, the orientation ratio is significantly reduced.

An important ingredient to obtain realistic screen images and orientation ratios is the alignment selectivity of the probe laser, which depends on its polarization [30]. Here, we consider a thermal sample in a laser pulse with $\tau = 10$ ns and $I_0 = 10^{12}\,W/cm^2$, and electric field $E_s = 300\,V/cm$ and $\beta = 30°$. In Fig. 8, we present its orientation ratio using the probe pulse with three possible polarizations. For a probe pulse linearly polarized parallel to the vertical axis of the screen, N_{up}/N_{tot} is the largest because such a pulse favors the measurement of the orientation on the screen. In contrast, if the probe pulse is linearly polarized perpendicular to the screen the orientation on the screen is reduced, and, therefore, N_{up}/N_{tot} presents the smallest values. The circularly polarized probe laser ensures that any molecule is ionized and detected with the same probability independently of the angle β, and provides the intermediate values of N_{up}/N_{tot} for any temperature. For a given state, there is no analytical relation between its orientation $\langle\langle\cos\theta\rangle\rangle_T$ and the orientation ratio N_{up}/N_{tot} of the 2D projection of its wave function, although the approximation $N_{up}/N_{tot} \approx (1 + \langle\langle\cos\theta\rangle\rangle_T)/2$ could be used to obtain an estimation. For instance, a 0.29 K thermal sample presents an orientation of $\langle\langle\cos\theta\rangle\rangle_T = 0.506$, and orientation ratios $N_{up}/N_{tot} = 0.757$ and 0.774 for a probe laser linearly polarized perpendicular and parallel to the screen detector, respectively, and $N_{up}/N_{tot} = 0.761$ for a circularly polarized one. These results should be compared with the value 0.753 given by this approach, which provides a lower bound for the N_{up}/N_{tot} of these three polarizations.

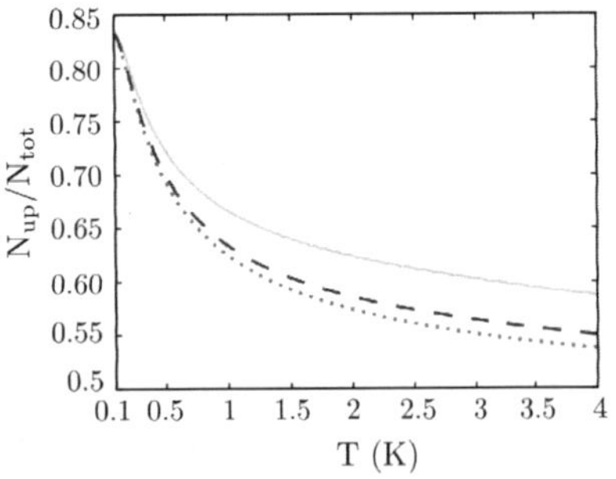

Figure 8. For a OCS thermal sample, we present the orientation ratio N_{up}/N_{tot} using a probe pulse linearly polarized along the vertical axis of the screen (thick solid line), along the perpendicular axis to the screen (dotted line) and circularly polarized in a plane perpendicular to the screen (dashed line). The field parameters are $\tau = 10$ ns, $I_0 = 10^{12}\,W/cm^2$, $E_s = 300\,V/cm$ and $\beta = 30°$.

For parallel fields, if the electric field strength is increased, the energy splitting in a pendular

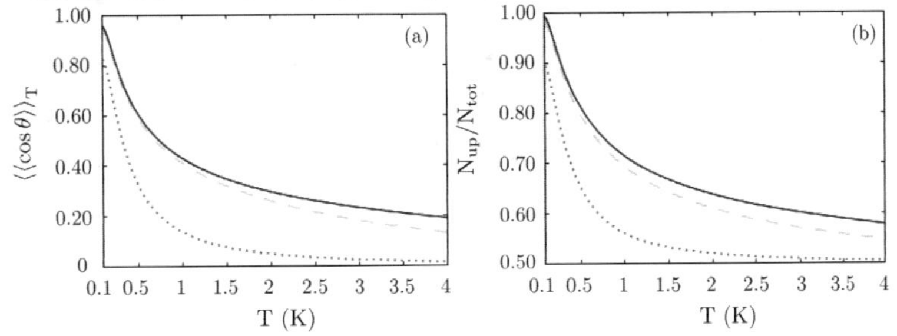

Figure 9. For a OCS thermal sample, we present (a) $\langle\langle\cos\theta\rangle\rangle_T$ and (b) N_{up}/N_{tot} as a function of the temperature for a 10 ns Gaussian pulse with peak intensities $I_0 = 10^{12}\,W/cm^2$ (thick solid line), $I_0 = 5\times10^{11}\,W/cm^2$ (dashed line) and $I_0 = 10^{11}\,W/cm^2$(dotted line). The field configuration is $E_s = 2\,kV/cm$ and $\beta = 30°$.

doublet is increased, and as a consequence, the degree of adiabaticity in the molecular mixed-field orientation is also enhanced. However, this statement only holds for the ground state of the two irreducible representations if the fields are tilted. For an excited rotational state, a strong dc field does not ensure a large orientation because the coupling between levels with different field-free M values becomes important, and this affects the molecular dynamics. In contrast, for a weak dc field, the mixing between these states is so small that M can be considered as conserved.

In Fig. 9, we plot $\langle\langle\cos\theta\rangle\rangle_T$ and N_{up}/N_{tot} for a thermal sample exposed to a 10 ns pulse combined with a dc field of $E_s = 2\,kV/cm$ tilted an angle $\beta = 30°$. For cold samples with $T \lesssim 0.74$ K and $T \lesssim 0.69$ K, we obtain $\langle\langle\cos\theta\rangle\rangle_T \gtrsim 0.5$ with $I_0 = 10^{12}\,W/cm^2$ and $5\times10^{11}\,W/cm^2$, respectively. For $I_0 = 5\times10^{11}\,W/cm^2$ and $10^{12}\,W/cm^2$, we obtain $N_{up}/N_{tot} \gtrsim 0.73$ if the rotational temperature is $T \lesssim 1.1$ K. Thus, using this strong dc field the orientation of a thermal ensemble becomes comparable to the experimental value for a quantum-state selected molecular beam in a very weak electric field. For this strong electric field, the orientation of the quantum-state selected beam is $N_{up}/N_{tot} = 0.99$ using a probe pulse linearly polarized along the vertical axis of the detector.

In Fig. 10, we present the expectation value $\langle\cos\theta\rangle_{JMl}$ at $t = 0$ for several rotational states in 10 ns Gaussian pulses with $I_0 = 5\times10^{11}\,W/cm^2$ and $10^{12}\,W/cm^2$, $E_s = 2\,kV/cm$ and $\beta = 30°$. For both field configurations, the $|0,0,e\rangle_0^{30}$ and $|3,1,e\rangle_0^{30}$ states are strongly oriented and antioriented, respectively. The remaining states show a moderate or weak orientation. The

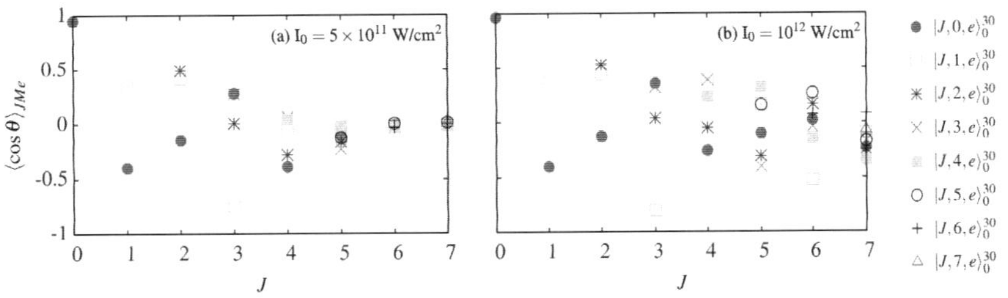

Figure 10. For the states $|J,M,e\rangle^{30}{}_0$, we present the orientation cosines $\langle\cos\theta\rangle_{JMl}$ at $t = 0$ versus the field-free rotational quantum number. The FWHM of the Gaussian pulses is $\tau = 10$ ns and the peak intensities (a) $I_0 = 5\times10^{11}\,W/cm^2$ and (b) $I_0 = 10^{12}\,W/cm^2$. The field configuration is $E_s = 2\,kV/cm$ and $\beta = 30°$.

effect of doubling the peak intensity is not noticeable for the levels with field-free rotational quantum number $J \leq 3$, and, in addition, for a certain peak intensity, we encounter similar orientation using a Gaussian pulse of 10 ns or 5 ns. The rotational dynamics of the ground state is adiabatic for both pulses; whereas for the excited state, this phenomenon can be explained by the non adiabatic effects taking place at weak laser intensities. When the levels in a certain J manifold are driven apart by the laser field, the process is nonadiabatic and there is a population transfer between them, already at weak laser intensities. Thus, the wave function of any excited level has contributions from adiabatic states which correspond to different pendular doublets. By further increasing the laser intensity, the molecular dynamics is affected by the avoided crossings with adjacent levels having different field-free magnetic quantum numbers M and by the formation of these pendular doublets. The rotational dynamics in most of these crossings will be nonadiabatic and has to be analyzed for each specific state. When the electric field is strong, the energy splitting within the states in the pendular pair is sufficiently large, and, as a consequence, the population transfer when the doublets are formed is not significant.

For completeness, in Fig. 6(b) we present the field-dressed rotational dynamics of the state $|2,0,e\rangle^{30}{}_t$ in a 10 ns pulse with $\mathrm{I}_0 = 10^{12}\,\mathrm{W/cm^2}$, a strong dc field of $\mathrm{E_s} = 2\,\mathrm{kV/cm}$ and $\beta = 30°$. After an adiabatic switching on of the electric field, the states in the $J = 2$ manifold are driven apart, $|C_{20e}(t)|^2$ decreases as $\mathrm{I}(t)$ is increased, whereas $|C_{21e}(t)|^2$ and $|C_{22e}(t)|^2$ increase. Compared to the weak dc field case in Fig. 6(a), this J-manifold splitting takes place at a stronger laser intensity, because the energy gap between the adiabatic states $|2,2,e\rangle^{30}_{\mathrm{p}}$, $|2,1,e\rangle^{30}_{\mathrm{p}}$ and $|2,0,e\rangle^{30}_{\mathrm{p}}$ is larger for $\mathrm{E_s} = 2\,\mathrm{kV/cm}$ than for $\mathrm{E_s} = 300\,\mathrm{V/cm}$. Let us mention that if $\mathrm{E_s}$ is increased, the energy splitting within this J-manifold is increased, and, therefore, the population redistribution will be reduced [16]. The avoided crossing between the states $|2,1,e\rangle^{30}_{\mathrm{p}}$ and $|2,2,e\rangle^{30}_{\mathrm{p}}$ occurs at $\mathrm{I}(t) \approx 2.96 \times 10^{10}\,\mathrm{W/cm^2}$, whereas the one involving the levels $|2,0,e\rangle^{30}_{\mathrm{p}}$ and $|3,3,e\rangle^{30}_{\mathrm{p}}$ around $\mathrm{I}(t) \approx 1.09 \times 10^{11}\,\mathrm{W/cm^2}$. Again, both of them are crossed diabatically, and the population of the adiabatic states is interchanged. By further increasing $\mathrm{I}(t)$, the pendular doublets start to form. In this case, the dc field is stronger and the energy gap is larger but the coupling due to the ac field is the same, then the population transfer is reduced. Indeed, the adiabatic states $|1,0,e\rangle^{30}_{\mathrm{p}}$, $|2,0,e\rangle^{30}_{\mathrm{p}}$ and $|3,2,e\rangle^{30}_{\mathrm{p}}$, the partners in the pendular doublets of $|2,2,e\rangle^{30}_{\mathrm{p}}$, $|2,1,e\rangle^{30}_{\mathrm{p}}$ and $|3,3,e\rangle^{30}_{\mathrm{p}}$, respectively, show a small population, which is below 0.01 once the peak intensity at $t = 0$ is achieved. Thus, the population at $t = 0$ for the field-dressed state $|2,0,e\rangle^{30}_{0}$ is $|C_{22e}(0)|^2 = 0.56$, $|C_{21e}(0)|^2 = 0.39$, and $|C_{33e}(0)|^2 = 0.05$. These results are similar for the four pulses formed by combining $\tau = 5$ ns and 10 ns with $\mathrm{I}_0 = 5 \times 10^{11}\,\mathrm{W/cm^2}$ and $10^{12}\,\mathrm{W/cm^2}$.

4. Conclusions

In this work, we investigate the mixed-field orientation dynamics of a thermal sample of linear molecules. We solve the time-dependent Schrödinger equation within the rigid rotor approximation for a large set of rotational states. As prototype example, we use the OCS molecule. However, we stress that the above results could be used to describe the mixed-field orientation of a thermal ensemble of other polar linear molecules by rescaling the Hamiltonian (1) in terms of the rotational constant.

By considering prototypical field configurations with weak dc fields, as in current mixed-field orientation experiments, we have proven that the rotational temperature of the molecular beam should be smaller than 0.7 K to achieve a significant orientation. Using a weak electric field, if the aim is a strongly oriented molecular ensemble, this should be as pure as possible in the ground state. Thus, it is required a quantum-state-selected molecular beam, unless the rotational temperature could be efficiently reduced below 1 K. It is found that a significant orientation is achieved for 1 K molecular samples when the electric field strength is increased.

Acknowledgments
Financial support by the Spanish project FIS2011-24540 (MICINN), the Grants P11-FQM-7276 and FQM-4643 (Junta de Andalucía), and the Andalusian research group FQM-207 is gratefully appreciated. J.J.O. acknowledges the support of ME under the program FPU.

References

[1] Brooks P R 1976 *Science* **193** 11
[2] Brooks P R and Jones E M 1966 *J. Chem. Phys.* **45** 3449
[3] Loesch H J and Möller J 1992 *J. Chem. Phys.* **97** 9016
[4] Aoiz F J, Friedrich B, Herrero V J, Rábanos V S and Verdasco J E 1998 *Chem. Phys. Lett.* **289** 132 – 140
[5] Aquilanti V, Bartolomei M, Pirani F, Cappelletti D and Vecchiocattivi F 2005 *Phys. Chem. Chem. Phys* **7** 291–300
[6] Bisgaard C Z, Clarkin O J, Wu G, Lee A M D, Gessner O, Hayden C C and Stolow A 2009 *Science* **323** 1464–1468
[7] Holmegaard L, Hansen J L, Kalhoj L, Kragh S L, Stapelfeldt H, Filsinger F, Küpper J, Meijer G, Dimitrovski D, Abu-samha M, Martiny C P J and Madsen L B 2010 *Nat. Phys.* **6** 428
[8] Hansen J L, Stapelfeldt H, Dimitrovski D, Abu-Samha M, Martiny C P J and Madsen L B 2011 *Phys. Rev. Lett.* **106** 073001
[9] Frumker E, Hebeisen C T, Kajumba N, Bertrand J B, Wörner H J, Spanner M, Villeneuve D M, Naumov A and Corkum P B 2012 *Phys. Rev. Lett.* **109** 113901
[10] Kraus P M, Vlajkovic S, Rupenyan A and Wörner H J 2012 *Phys. Rev. Lett.* Accepted
[11] Spector L S, Artamonov M, Miyabe S, Martinez T, Seideman T, Guehr M and Bucksbaum P H 2012 ArXiv:1207.2517
[12] Gijsbertsen A, Siu W, Kling M F, Johnsson P, Jansen P, Stolte S and Vrakking M J J 2007 *Phys. Rev. Lett.* **99** 213003
[13] Friedrich B and Herschbach D R 1999 *J. Chem. Phys.* **111** 6157
[14] Friedrich B and Herschbach D 1999 *J. Phys. Chem. A* **103** 10280
[15] Nielsen J H, Stapelfeldt H, Küpper J, Friedrich B, Omiste J J and González-Férez R 2012 *Phys. Rev. Lett.* **108** 193001
[16] Omiste J J and González-Férez R 2012 *Phys. Rev. A* **86** 043437
[17] Sakai H, Minemoto S, Nanjo H, Tanji H and Suzuki T 2003 *Phys. Rev. Lett.* **90** 083001
[18] Buck U and Fárník M 2006 *Int. Rev. Phys. Chem.* **25** 583
[19] Ghafur O, Rouzee A, Gijsbertsen A, Siu W K, Stolte S and Vrakking M J J 2009 *Nat Phys* **5** 289–293
[20] Holmegaard L, Nielsen J H, Nevo I, Stapelfeldt H, Filsinger F, Küpper J and Meijer G 2009 *Phys. Rev. Lett.* **102** 023001
[21] Filsinger F, Küpper J, Meijer G, Holmegaard L, Nielsen J H, Nevo I, Hansen J L and Stapelfeldt H 2009 *J. Chem. Phys.* **131** 064309
[22] Even U, Jortner J, Noy D, Lavie N and Cossart-Magos C 2000 *J. Chem. Phys.* **112** 8068
[23] Seideman T and Hamilton E 2006 *Adv. Atom. Mol. Opt. Phys.* **52** 289
[24] Feit M D, Fleck Jr J A and Steiger A 1982 *J. Comp. Phys.* **47** 412
[25] Bačič Z and Light J C 1989 *Annu. Rev. Phys. Chem.* **40** 469
[26] Corey G C and Lemoine D 1992 *J. Chem. Phys.* **97** 4115
[27] Offer A R and Balint-Kurti G G 1994 *J. Chem. Phys.* **101** 10416–10428
[28] Sánchez-Moreno P, González-Férez R and Schmelcher P 2007 *Phys. Rev. A* **76** 053413
[29] Härtelt M and Friedrich B 2008 *J. Chem. Phys.* **128** 224313
[30] Omiste J J, Gärttner M, Schmelcher P, González-Férez R, Holmegaard L, Nielsen J H, Stapelfeldt H and Küpper J 2011 *Phys. Chem. Chem. Phys.* **13** 18815–18824

Journal of Physics: Conference Series **467** (2013) 012006 doi:10.1088/1742-6596/467/1/012006

Two-photon laser spectroscopy of antiprotonic helium atoms, and the antiproton-to-electron mass ratio

Masaki Hori

Max-Planck-Institut für Quantenoptik, Hans-Kopfermann-Strasse 1, 85748 Garching, Germany

E-mail: Masaki.Hori@mpq.mpg.de

Abstract. Some two-photon transitions in antiprotonic helium atoms at the deep UV wavelengths λ = 139.8–197.0 nm were recently studied by laser spectroscopy. The thermal Doppler broadening of the observed antiprotonic resonances were reduced by exciting the atoms with two counterpropagating laser beams of wavelengths λ = 265–417 nm. The resulting narrow spectral lines allowed the measurement of three transition frequencies in antiprotonic helium-3 and helium-4 isotopes with fractional precisions of 2.3–5 parts in 10^9. By comparing the results with three-body QED calculations, the antiproton-to-electron mass ratio was derived as 1836.1526736(23). We briefly review these experimental results that were presented in Ref. [1].

1. Introduction

Metastable antiprotonic helium ($\bar{p}He^+$) is a three-body atom [1, 2, 3, 4] consisting of a helium nucleus, an electron in the 1s state, and an antiproton occupying a Rydberg state with high principal and angular momentum quantum numbers $n \sim \ell + 1 \sim 38$. The atom retains microsecond-scale lifetimes against antiproton annihilation in the helium nucleus, due to the fact that the antiproton orbital has a negligible overlap with the nucleus; the 1s electron protects the antiproton against collisions with other helium atoms. This longevity makes $\bar{p}He^+$ especially amenable to precision laser spectroscopy [5, 6, 7]. The atomic transition frequencies of $\bar{p}He^+$ have been calculated by three-body QED calculations to fractional precisions of 1×10^{-9} [8]. These calculations included relativistic and radiative recoil corrections up to order $m_e c^2 \alpha^6 / h$, and nuclear size effects. By comparing the measured and calculated transition frequencies, the antiproton-to-electron mass ratio can be determined to parts-per-billion scale precision.

The ASACUSA collaboration at CERN [9] has measured some $\bar{p}He^+$ transition frequencies in the optical region $0.3 - 1$ PHz with a fractional precision of $10^{-7} - 10^{-8}$, by single-photon laser spectroscopy [5, 6, 7]. The precision was limited by the Doppler broadening of the resonance lines which arose from the thermal motions of the $\bar{p}He^+$ in the target. Unlike the atomic hydrogen case [10], it is difficult to cancel the first-order Doppler broadening by irradiating the atom with two equal-frequency photons and inducing a two-photon transition; the probabilities involved in these nonlinear transitions of the massive antiproton are too small (a few a.u.). In fact, calculations indicate that gigawatt-scale laser powers would be needed to excite the antiproton within the atom's microsecond-scale lifetime against annihilation [11].

In a recent experiment, two-photon transitions of the type $(n, \ell) = (n-2, \ell-2)$ [1] [Fig. 1(a)] were excited by utilizing the fact that the probability can be enhanced by factor 10^4, to

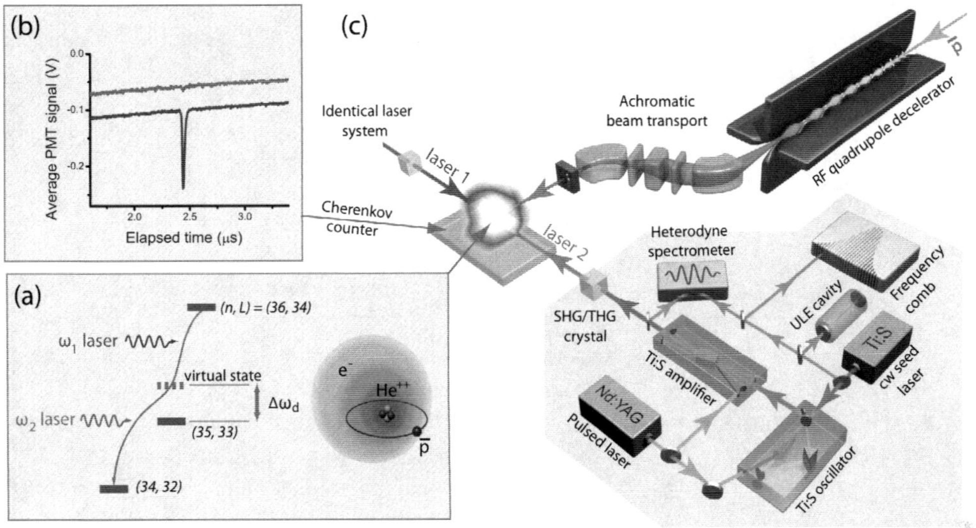

Figure 1. Schematic energy level diagram of $\overline{p}^4He^+$ states involved in the two-photon transition $(n, \ell) = (36, 34) \rightarrow (34, 32)$ (a). The virtual intermediate state is tuned some $\Delta\omega_c \sim 10$ GHz from a real state $(35, 33)$. Cherenkov detector signals for two-photon transition (b). Experimental layout (c). See Ref. [1] for details.

around $\sim 10^4$ a.u., if the counterpropagating beams have non-equal frequencies ν_1 and ν_2, such that the virtual intermediate state of the two-photon transition lies within a few GHz of a real $\overline{p}He^+$ state $(n-1, \ell-1)$. The first-order Doppler width in the observed resonance lines can then be reduced by a factor $|\nu_1 - \nu_2|/(\nu_1 + \nu_2) \sim 1/20$.

2. Experiment

The two-photon transitions were induced between $\overline{p}He^+$ states with microsecond and nanosecond-scale lifetimes against Auger emission of the electron. After Auger decay, the remaining two-body $\overline{p}He^{2+}$ ion [12] was rapidly destroyed by Stark collisions with other helium atoms in the experimental target. The charged pions emerging from the resulting antiproton annihilations were detected by Cherenkov detectors [13] placed around the experimental target. The two-photon resonance condition between the counterpropagating laser beams and the $\overline{p}He^+$ was thus revealed as a sharp spike in the rate of antiproton annihilations [Fig. 1(b)].

Megawatt-scale laser pulses of high spectral purity are needed to excite these nonlinear two-photon transitions, that have amplitudes of 10^3–10^4 a.u. We therefore developed two sets of Ti:Sapphire lasers [14] of pulse length 30-100 ns with a narrow spectral linewidth (~ 6 MHz). The laser system included continuous-wave (cw) lasers of wavelengths 728–940 nm, whose frequencies were measured to a precision of $< 1 \times 10^{-10}$ against a femtosecond optical frequency comb [15]. This beam was then used to seed a ring Ti:Sapphire oscillator and multipass amplifier which generated laser pulses of energy 50–100 mJ.

Nanosecond-scale changes in the refractive index of the Ti:Sapphire crystals during the amplification, as well as the so-called "mode-pulling" effects in the pulsed laser cavity [14], caused the laser linewidth to broaden and the frequency to shift by several tens MHz. This frequency chirp was measured using a heterodyne spectrometer, and corrected by intracavity electro-optic modulators located inside the Ti:Sapphire cavity. The spectral precision ($< 1.4 \times 10^{-9}$) of the

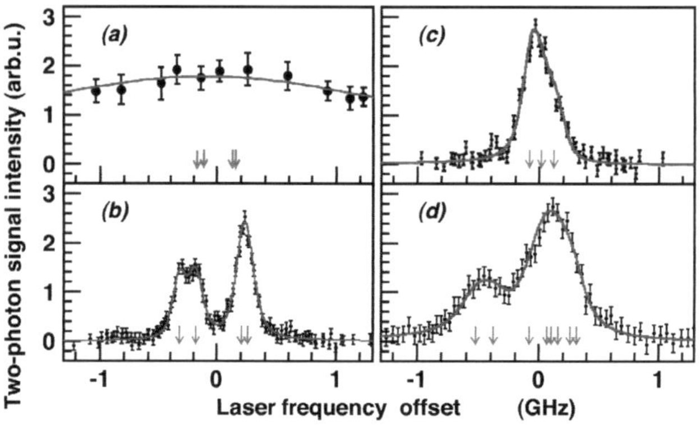

Figure 2. Doppler- and power-broadened profile of the single-photon resonance $(36, 34) \rightarrow (35, 33)$ of $\overline{p}^4He^+$ (a). Sub-Doppler two-photon profiles of $(36, 34) \rightarrow (34, 32)$ (b) and $(33, 32) \rightarrow (31, 30)$ (c) of $\overline{p}^4He^+$, and $(35, 33) \rightarrow (33, 31)$ of $\overline{p}^3He^+$ (d). Solid lines indicate best fit of theoretical line profiles (see text) and partly overlapping arrows the positions of the hyperfine lines. From Ref. [1].

pulsed laser was verified by measuring some two-photon transition frequencies in Rb and Cs at wavelengths of 778 and 822 nm.

The Antiproton Decelerator (AD) facility of CERN provided 200-ns-long pulsed beams, which contained $\sim 10^7$ antiprotons of kinetic energy 5.3 MeV, at a repetition rate of 0.01 Hz [Fig. 1 (c)]. The antiprotons were decelerated to ~ 70 keV, by allowing them to pass through a 3-m-long radiofrequency quadrupole decelerator [6]. Secondary electron emission detectors measured the spatial profiles of the beam [16]. The antiprotons were allowed to stop in a cryogenic target chamber filled with ^4He or ^3He gas at temperature $T \sim 15$ K and pressure $p = 0.8 - 3$ mbar. Two, horizontally-polarized laser beams of energy density ~ 1 mJ/cm^2 fired through the target in a perpendicular direction to the antiproton beam excited the two-photon transitions.

The Cherenkov signal corresponding to some 10^7 $\overline{p}He^+$ atoms is shown in Fig. 1(b), as a function of time elapsed since the arrival of antiproton pulses at the experimental target. Lasers of wavelengths $c/\nu_1 = 417$ and $c/\nu_2 = 372$ nm were tuned to the two-photon transition $(n, \ell) = (36, 34) \rightarrow (34, 32)$, so that the virtual intermediate state lay $\Delta\nu_d \sim 6$ GHz away from the real state $(35, 33)$. The annihilation spike which corresponds to the two-photon transition can be seen at $t = 2.4\mu$s. The intensity of the spike reflects the number of antiprotons populating state $(36, 34)$ [17, 18, 19]. When the 417-nm laser was tuned some ~ 0.5 GHz off the two-photon resonance condition, the signal disappeared as indicated in the same figure.

Fig. 2(b) shows the resonance profile which was measured by detuning the ν_1 laser to $\Delta\nu_d = -6$ GHz, whereas the ν_2 laser was scanned between -1 and 1 GHz around the two-photon resonance defined by $\nu_1 + \nu_2$. The linewidth (~ 200 MHz) of this two-photon resonance is more than an order of magnitude smaller than the Doppler- and power-broadened profile of the single-photon one $(36, 34) \rightarrow (35, 33)$ [Fig. 2(a)]. The two-peak structure, with a frequency interval of 500 MHz, arises due to the interaction between the electron spin and the orbital angular momentum of the antiproton. We also detected the $(33, 32) \rightarrow (31, 30)$ resonance of $\overline{p}^4He^+$ at a wavelength of 139.8 nm, using lasers of $c/\nu_1 = 296$ nm and $c/\nu_2 = 265$ nm [Fig. 2(c)] detuned $\Delta\nu_d \sim 3$ GHz from state $(32, 31)$. All four hyperfine lines are much closer together (± 100 MHz). We also measured the $\overline{p}^3He^+$ resonance $(35, 33) \rightarrow (33, 31)$ of $\lambda = 193.0$ nm [Fig. 2(d)] using lasers of 410 and 364 nm. This resonance profile contains eight partially-overlapping hyperfine

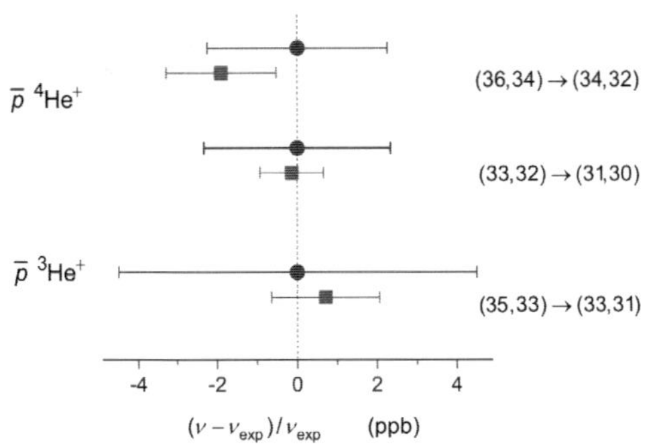

Figure 3. Fractional deviation between theoretical (squares) and experimental (circles) transition frequencies of $\bar{p}$He$^+$ isotopes measured by two-photon laser spectroscopy. From Ref. [1].

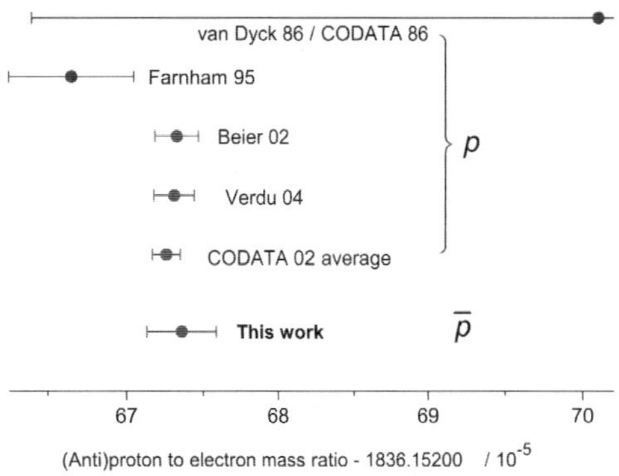

Figure 4. Antiproton-to-electron mass ratio determined in this work, compared with the proton-to-electron mass ratios measured in Penning trap experiments [22, 23, 24, 25] and the CODATA 2002 recommended value obtained by averaging them. From Ref. [1].

lines, which arose from the spin-spin interactions of the three constituent particles.

The spin-independent transition frequencies ν_{exp} were obtained by fitting these measured profiles with a theoretical lineshape (solid lines in Fig. 2) which was determined by numerically solving the rate equations of the two-photon process [11]. This included the transition rates, power and Doppler broadening effects, frequency modulation in the laser pulse, the experimentally-measured spatial and temporal profiles of the laser beam, and ac Stark effects. The positions of the hyperfine lines were fixed to the theoretical values calculated by Korobov [20], which have a precision of < 0.5 MHz. For the transition $(36, 34) \to (34, 32)$ in $\bar{p}^4$He$^+$, the statistical uncertainty due to the finite number of atoms in the laser beam was estimated as 3 MHz. For the target densities studied here $\rho = (1-3) \times 10^{18}$ cm^{-3}, no significant collisional shift was observable within the experimental error. This agrees with quantum chemistry calculations [21], for which the predictions of 0.1–1-MHz collisional shifts in the single-photon lines agreed with experimental results [7] with a precision of $\leq 20\%$. Theoretical calculations [11] also show that magnetic Zeeman shifts are also small < 0.5 MHz. The ac Stark shift [11] was reduced to ≤ 5 MHz by adjusting the relative intensities of the two laser beams. Remaining ac Stark shifts were canceled to a level of 0.5MHz by systematically comparing the resonance profiles measured at positive and negative detunings $\pm\Delta\nu_d$ of the virtual intermediate state. The experimental uncertainty σ_{exp} was obtained as the quadratic sum of all these errors.

3. Results and conclusions

The experimental transition frequencies ν_{exp} (filled circles with error bars in Fig. 3) agree with the QED calculated ν_{th} values (squares) within a fractional precision of $(2-5) \times 10^{-9}$. The calculation uses the fundamental constants compiled in CODATA2002 [22], such as the ^{3}He- and ^{4}He-to-electron mass ratios, the Bohr radius, and Rydberg constant. The charge radii of the ^{3}He and ^{4}He nuclei give relatively small corrections to ν_{th} of $4-7$ MHz, whereas the correction from the antiproton radius is less than 1 MHz. The small contributions are due to the fact that the wavefunctions of the antiprotonic states with large ℓ-value have only a negligible overlap with the helium nucleus. The theoretical precision of $\nu_{\rm th}$ is now mainly limited by the uncalculated radiative corrections of order $m_e c^2 \alpha^8/h$ [8], but higher-order corrections are now being calculated. When the antiproton-to-electron mass ratio $M_{\overline{p}}/m_e$ in these calculations was increased by a relative amount of 10^{-9}, the $\nu_{\rm th}$-value changed by 2.3–2.8 MHz. By minimizing the difference between $\nu_{\rm th}$ and $\nu_{\rm exp}$ and considering the systematic errors, we obtained the antiproton-to-electron mass ratio as,

$$M_{\overline{p}}/m_e = 1836.1526736(23), \tag{1}$$

which yielded the best agreement between theoretical and experimental frequencies. The uncertainty includes the statistical and systematic experimental, and theoretical contributions of 18×10^{-7}, 12×10^{-7}, and 10×10^{-7}. This is in good agreement with previous measurements[22, 23, 24, 25] of the proton-to-electron mass ratio (Fig. 4), which have a similar experimental precision. By assuming that CPT invariance is valid (i.e, $M_{\overline{p}} = M_p = 1.00727646677(10)$ u), we further derived a value for the electron mass, $m_e = 0.0005485799091(7)$ u, from this experimental result on $\overline{p}He^+$ [1].

Hughes and Deutch [26, 27] has constrained any difference between the antiproton and proton charges and masses $\delta_Q = (Q_p + Q_{\overline{p}})/Q_p$ and $\delta_M = (M_p - M_{\overline{p}})/M_p$ to better than 2×10^{-5}. To do this, they combined X-ray spectroscopic data of antiprotonic atoms (proportional to $Q_{\overline{p}}^2 M_{\overline{p}}$) and the cyclotron frequency ($\propto Q_{\overline{p}}/M_{\overline{p}}$) of antiprotons confined in magnetic Penning traps measured to a higher precision. We improved this limit by studying the linear dependence of δ_M and δ_Q on $\nu_{\rm th}$ of $\overline{p}He^+$, i.e., $\delta_M \kappa_M + \delta_Q \kappa_Q < |\nu_{\rm exp} - \nu_{\rm th}|/\nu_{\rm exp}$ [2]. For the three $\overline{p}He^+$ transitions studied in this work, the constants were estimated as $\kappa_M = 2.3 - 2.8$ and $\kappa_Q = 2.7 - 3.4$, whereas the right side of this equation was evaluated by averaging over the three transitions as $< (8 \pm 15) \times 10^{-10}$. Meanwhile the constraint of $(Q_{\overline{p}}/M_{\overline{p}})/(Q_p/M_p) + 1 = 1.6(9) \times 10^{-10}$ from the TRAP experiment [28, 29] implies that $\delta_Q \sim \delta_M$. We conclude from this that any deviation between the charges and masses of protons and antiprotons are less than 7×10^{-10} at 90% confidence level [1].

We are currently attempting to further improve the experimental and theoretical uncertainties in these experiments, by e.g., cooling the atoms to lower temperature and improving the quality of the antiproton beam.

Table 1. Spin-averaged transition frequencies of $\overline{p}He^+$. Experimental values show respective total, statistical and systematic errors in parentheses. Theoretical values (From Ref. [8] and V. I. Korobov, private communication) show respective uncertainties from uncalculated QED terms and numerical errors in parentheses. From Ref. [1].

Isotope	Transition $(n, \ell) \to (n-2, \ell-2)$	Transition frequency (MHz) Experiment	Theory
$\overline{p}^4He^+$	$(36, 34) \to (34, 32)$	1,522,107,062(4)(3)(2)	1,522,107,058.9(2.1)(0.3)
	$(33, 32) \to (31, 30)$	2,145,054,858(5)(5)(2)	2,145,054,857.9(1.6)(0.3)
$\overline{p}^3He^+$	$(35, 33) \to (33, 31)$	1,553,643,100(7)(7)(3)	1,553,643,100.7(2.2)(0.2)

Acknowledgments

This work was supported by the European Research Council (ERC-Stg), Monbukagakusho (grant no 20002003), Hungarian Research Foundation (K72172), and the Austrian Federal Ministry of

Science and Research. I thank the ASACUSA collaborators, CERN AD and PS operational staff, the CERN cryogenics laboratory, J. Alnis, D. Bakalov, J. Eades, R. Holzwarth, V.I. Korobov, M. Mitani, W. Pirkl, and T. Udem.

References

[1] Hori M *et al.* 2011 *Nature* **475** 484
[2] Hayano R S, Hori M, Horváth D and Widmann E 2007 *Rep. Prog. Phys.* **70** 1995
[3] Condo G T 1964 *Phys. Lett.* **9** 65
[4] Russel J E 1969 *Phys. Rev. Lett.* **23** 63
[5] Hori M *et al.* 2001 *Phys. Rev. Lett.* **87** 093401
[6] Hori M *et al.* 2003 *Phys. Rev. Lett.* **91** 123401
[7] Hori M *et al.* 2006 *Phys. Rev. Lett.* **96** 243401
[8] Korobov V I 2008 *Phys. Rev. A* **77** 042506
[9] Hori M and Walz J 2013 *Progress in Particle and Nuclear Physics* doi:10.1016/j.ppnp.2013.02.004
[10] Parthey Ch G *et al.* 2011 *Phys. Rev. Lett.* **107** 203001
[11] Hori M and Korobov V I 2010 *Phys. Rev. A* **81** 062508
[12] Hori M *et al.* 2005 *Phys. Rev. Lett.* **94** 063401
[13] Hori M *et al.* 2003 *Nucl. Instrum. Methods Phys. Res. A* **496** 102
[14] Hori M and Dax A 2009 *Opt. Lett.* **34** 1273
[15] Udem Th, Holzwarth R and Hänsch T W 2002 *Nature* **416** 233
[16] Hori M 2005 *Rev. Sci. Instrum.* **76** 113303
[17] Hori M *et al.* 2002 *Phys. Rev. Lett.* **89** 093401
[18] Hori M *et al.* 2004 *Phys. Rev. A* **70** 012504
[19] Hori M *et al.* 1998 *Phys. Rev. A* **57** 1698; 1998 *Phys. Rev. A* **58** 1612
[20] Korobov V I 2006 *Phys. Rev. A* **73** 022509
[21] Bakalov D *et al.* 2000 *Phys. Rev. Lett.* **84** 235
[22] Mohr P J, Taylor B N and Newell D B 2008 *Rev. Mod. Phys.* **80** 633
[23] Farnham D L, Van Dyck Jr. R S, Schwinberg P B 1995 *Phys. Rev. Lett.* **75** 3598
[24] Beier T *et al.* 2002 *Phys. Rev. Lett.* **88** 011603
[25] Verdú J *et al.* 2004 *Phys. Rev. Lett.* **92** 093002
[26] Hughes R J and Deutch B I 1992 *Phys. Rev. Lett.* **69** 578
[27] Nakamura K *et al.* 2010 *J. Phys.* **G37** 075021
[28] Gabrielse G *et al.* 1999 *Phys. Rev. Lett.* **82** 3198
[29] Thompson J K, Rainville S and Pritchard D E 2004 *Nature* **430** 58

Quantum noise in optical tweezers

Michael A. Taylor and Warwick P. Bowen

Centre for Engineered Quantum Systems, University of Queensland, St Lucia, Queensland 4072, Australia

E-mail: `wbowen@physics.uq.edu.au`

Abstract. Quantum enhanced sensitivity in optical tweezers based particle tracking was recently demonstrated. This has provided the necessary tool for quantum metrology to play an important role in biological measurements. Here we introduce the basic theory relevant to such optical tweezers experiments, and overview the significance of sub-shot noise limited sensitivity to practical experiments. In particular, biophysical experiments are subject to optical power constraints, which therefore limits the absolute sensitivity which is classically achievable. Quantum enhanced particle tracking can overcome this limit, and is therefore likely to play an important role in such biophysical experiments in the near future.

1. Introduction

The development of optical tweezers made it possible to manipulate and measure the motion of single microscopic particles [1]. This revolutionized the field of biophysics by allowing the study of single-molecule kinetics, without need for ensemble averaging [2, 3]. For instance, this advance enabled observation of steps in the motion of the protein motor kinesin [4] and the muscle protein myosin [5]. It also allows direct measurement of the stretching and folding properties of DNA and RNA [3, 6, 7], the strain on an enzyme during catalysis [6], and the dynamics of virus-host coupling [8].

After decades of research, optical tweezers continue to advance technologically, which allows observation of an ever-increasing range of biophysical phenomena [9]. With these advances, leading experiments are approaching the quantum limit on sensitivity, such that further improvements will soon require integration of quantum correlated light [10]. Recently, the first optical tweezers experiment with sensitivity beyond the quantum shot noise limit was demonstrated [11]. Here we briefly overview the significance and outlook of such quantum techniques in optical tweezers.

2. Benefit of quantum enhanced sensitivity

In any optical measurement, quantum noise sets a fundamental limit on the achievable sensitivity [12, 13]. This limit can only be surpassed if quantum correlations between photons are used to suppress the quantum noise, such that more more information is extracted per photon [14, 15]. Hence, quantum metrology allows high sensitivity measurements to proceed with a lower light intensity than classically possible [16, 17], and in the presence of optical power constraints, allows sensitivity which can outperform any classical competitor. Of the quantum technologies available, squeezed light is the most promising technology for practical measurements, as it is compatible with high photon flux and offers the best possible performance in the presence of optical losses [15]. In most experiments, however, it is easier to simply increase the power than to integrate quantum correlated light. The only practical application of quantum metrology so far has been the use of squeezed light in gravity wave observatories, where the optical power already operates near the damage threshold of the mirrors [18]. Another frontier where quantum metrology could be important is in biological measurements [16, 17, 19], where observation of nanoscale features classically requires a high photon flux that also introduces disruptive photochemical interactions [20, 21].

In biological measurements, any incident light interacts strongly with living cells. Laser heating increases the temperature and induces temperature gradients, which can cause vast changes to the cellular respiration and division, and can also destroy the cell [20, 22]. This heating effect is minimized by using light in the visible spectrum, where water is highly transparent. However, the light also directly writes chemical changes into living cells by producing reactive oxygen species [23]. It is known that the cell division rate is substantially changed by illumination [24], as is enzyme activity [25] and many other processes. Photochemical interactions are particularly well studied in nerve cells, where they allow controlled neural stimulation [26]. When too much light is used, these photochemical effects are fatal for the cells [21]. The damaging chemical effects are dramatically reduced by using infra-red laser beams [2]. For optical tweezers experiments, 1064 nm lasers are a very common choice, as these lasers are low-noise and this wavelength approximately minimizes the collective photochemical and photothermal cellular damage [22, 2]. However, while minimizing the damage can make the light non-fatal to the cell, it still perturbs a wide range of cell functions, which can influence the biological parameter under study.

3. Shot-noise in optical tweezers

In many optical tweezers experiments, a biological molecule exerts a force on a trapped particle, and the particle position is measured to determine the biomolecular force. Since such experiments are performed in an aqueous environment at room temperature, the desired force competes with the stochastic thermal force on the particle. Force resolution is therefore limited by both the absolute sensitivity of the apparatus and the background of thermal forces. These experiments are typically limited by detection drifts at sub-Hz frequencies, thermal noise over the bulk of the detection range, and are shot-noise limited only above a few tens of kHz [27]. While providing additional measurement bandwidth, sub-shot noise sensitivity would not generally improve the signal-to-noise ratio at biologically relevant frequencies in the Hz range. Additionally, such experiments rely on optical forces which scale with the light intensity, such that quantum enhanced sensitivity alone does not necessarily allow the use of lower photon flux.

Although thermal fluctuations can limit the resolution of non-thermal forces, they also provide a method to probe the mechanical nature of the fluid surrounding the particle. The thermal motion of a particle is determined both by the thermal force and the mechanical properties of its surrounding medium. Hence, particle tracking allows characterization of the viscosity and elasticity of the surrounding fluid [28, 29], as well as the hydrodynamic properties of the system which can couple the particle motion to surrounding walls [30, 31] and other particles [32], and also introduces non-Markovian memory to the viscous damping [33, 34, 35]. These characteristics are all extremely important in cellular systems, since they influence all chemical processes which are mediated by thermal motion. The sensitivity of such experiments is limited by detection noise at all frequencies, so quantum enhanced sensitivity allows characterization of such effects with improved bandwidth and precision. Furthermore, optical forces are not typically required in such experiments, so enhanced sensitivity allows reduced photon flux and a corresponding reduction in photochemical interactions.

4. Optical trapping theory

Optical manipulation in optical tweezers is based on the interaction between a tightly focused laser beam and a small particle (see Fig. 1*a*). The momentum of the optical field changes as the particle scatters light, which thus imparts a force on the particle. The resulting force is simple to evaluate if the trapped particle is approximated as a point dipole [36]. In this case, the optical field $\mathbf{E}$ induces an oscillating dipole with polarization $\mathbf{p} = \alpha\mathbf{E}$, where α is the polarizability. Neglecting components at optical frequencies, the optical gradient force on the dipole is given by

$$\mathbf{F}_{\text{grad}} = (\mathbf{p}\cdot\nabla)\,\mathbf{E} = \frac{1}{2}\alpha\,\nabla\,E^2, \tag{1}$$

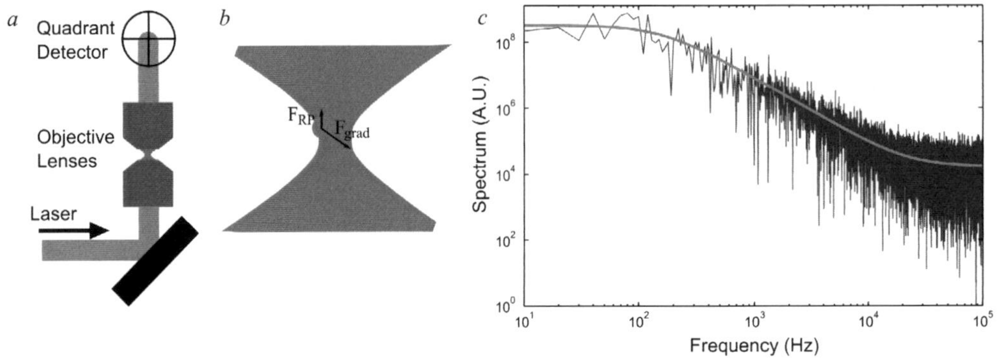

Figure 1. (*a*) The most basic elements of an optical tweezers experiment. Laser light is focused to a point to trap particles, and then collected for measurement. In most cases, this measurement occurs on a quadrant detector as shown here. (*b*) The particle is trapped near the focus of the beam, and is subject to two forces; a gradient force F_{grad}, which points toward the intensity maxima (see Eq. 1), and radiation pressure F_{RP} which pushes in the direction of light propagation (see Eq. 2). (*c*) A spectra as is typical for trapped Brownian motion. The expectation value is given by a Lorentzian mechanical profile (Eq. 5) along with a noise floor.

where the vector identity $\nabla E^2 = 2(\mathbf{E}\cdot\nabla)\mathbf{E} + 2\mathbf{E}\times(\nabla\times\mathbf{E})$ has been used, along with $\nabla\times\mathbf{E} = 0$ from Maxwell's equations [36]. The force described in Eq. 1 points toward maxima in intensity for particles with positive polarizability, or equivalently, a higher refractive index than the surrounding medium. In addition to this force, the radiation pressure of the trapping laser also pushes the particle along the optical axis $\hat{z}$, which tends to destabilize the trap (see Fig. 1*b*). This force is given by

$$\mathbf{F}_{\text{RP}} = \frac{k^4\alpha^2}{12\pi n^2\epsilon_0}E^2\hat{z}, \tag{2}$$

where k is the wavenumber, n the refractive index of the surrounding medium, and ϵ_0 the vacuum permittivity [36]. To enable stable trapping, the trapping force must dominate the radiation pressure. To achieve this, the intensity gradient is maximized by focusing the laser to the smallest spot possible. Additionally, the scattering rate is important because the gradient force scales as α while the radiation pressure scales as α^2. The scattering rate therefore should not be vanishingly small, as this produces minimal optical forces, but it also cannot be very large, as this produces large radiation pressure forces which destabilize the optical trap. Both the gradient and radiation pressure optical forces scale linearly with the light intensity, so there is no upper limit on the optical power which can be used to trap particles.

When an optical force is exerted on the particle, the momentum of the light changes such that the laser beam is deflected. Thus, measurement of the direction of the transmitted light provides a simple and effective method to measure the particle position. This measurement is typically performed by collecting the transmitted light with an objective lens and directing it onto a quadrant detector [37].

5. Trapped Brownian motion spectra

Here we derive the expected thermal trajectory $x(t)$ of a trapped particle in the simple case of a purely viscous medium with no elasticity or hydrodynamic interactions. In this case, the equation of motion along one dimension is given by

$$m\ddot{x}(t) = -\gamma\dot{x}(t) - k_{\text{opt}}x(t) + F_T, \tag{3}$$

for a particle mass m, friction coefficient γ, a harmonic optical force $-k_{\rm opt}x(t)$, and a fluctuating thermal force $F_T = (2k_BT\gamma)^{1/2}\xi(t)$, with T the temperature and k_B Boltzmann's constant [38]. The fluctuating function has the properties $\langle\xi(t)\rangle = 0$ and $\langle\xi(t)\xi(t')\rangle = \delta(t-t')$. By taking the Fourier transform of Eq. 3 and rearranging, we find that

$$x(\omega) = \frac{(2k_BT\gamma)^{1/2}}{-m\omega^2 + i\omega\gamma + k_{\rm opt}}\xi(\omega). \quad (4)$$

From this we can see that the inertial term $m\ddot{x}(t)$ is negligible at frequencies far below the rate at which the kinetic energy is lost through friction, $\omega \ll \gamma/m$. The friction is given by Stokes' law as $\gamma = 6\pi\eta R$, where R is the particle radius and η is the fluid viscosity. The inertial contribution is measurable only at very short timescales [39, 40]; for instance, a 1 μm diameter silica microsphere in water has m/γ = 120 ns. This term is therefore typically neglected [38], with the resulting power spectra of trapped Brownian motion given by

$$\langle|x(\omega)|^2\rangle = \frac{2k_BT/\gamma}{\omega^2 + \omega_c^2}, \quad (5)$$

where we have used $\langle|\xi(\omega)^2|\rangle = 1$ and introduced the corner frequency $\omega_c \equiv k_{\rm opt}/\gamma$. Measurements of this mechanical spectra will also include a detection noise floor, as shown in Fig. 1*c*. This Lorentzian spectral shape is characteristic of trapped Brownian motion. In many regimes, particularly in biological systems, the spectral shape is altered by fluid elasticity or hydrodynamic interactions. These effects can be characterized by determining the profile of the thermal motion. If squeezed light is used, the minimum resolvable displacement is smaller and the profile of thermal motion can be characterized more precisely and over a wider bandwidth. Alternatively, equivalent sensitivity is achievable with reduced light intensity, which therefore reduces damage to the specimen.

6. Status of the field

In Ref. [11], the quantum shot noise limit was overcome in optical tweezers based particle tracking, and then quantum enhanced sensitivity applied in measurements of living systems. In that work, a microparticle tracking technique was developed which is both compatible with non-classical light and immune to low frequency optical and electronic noise which would otherwise dominate shot-noise at the Hz–kHz frequencies of interest [41]. For the first time, this provided the tools necessary to perform a broad range of quantum enhanced measurements in biology.

This apparatus was then applied to tracking the thermal motion of naturally occurring lipid granules within *Saccharomyces cerevisiae* yeast cells, with amplitude squeezed light yielding a 2.4 dB enhancement in displacement sensitivity. Since the cellular cytoplasm is crowded with large molecules, polymer networks, and various other organelles, the thermal motion of embedded particles differs from the Brownian motion described above. This thermal motion was characterized with quantum enhanced sensitivity, thus allowing enhanced characterization of the mechanical properties of the cellular cytoplasm, and allowing dynamic changes in these properties to be measured with enhanced temporal resolution.

Following this demonstration, the biological experiment was extended to spatially resolved sub-diffraction limited quantum imaging of the cytoplasmic structure [42]. Spatial variations in the local mechanical properties of the cellular cytoplasm were sampled by lipid granules as they diffused thermally through the cells. These spatial structures were resolved at length scales down to 10 nm, far below the diffraction limit. In this case the use of quantum correlated light enhanced the spatial resolution by 14%. If combined with state-of-the-art squeezed light with over 10 dB of measured squeezing [43], this technique is predicted to allow up to an order of magnitude improvement in resolution over similar classical imaging techniques.

These experiments firmly demonstrated that sub-shot noise limited biological measurements are possible in optical tweezers. However, this was demonstrated in an apparatus which used microscope objectives with a relatively low numerical aperture (NA) of 0.4, such that the collection angle was smaller than is typically used in similar classical experiments. Although the absolute sensitivity was comparable to that achieved in similar classical experiments [44], the experiment was not classically optimal, and could be improved substantially by integrating higher NA objectives. However, microscope objectives are generally designed for operation in the visible, while the optical tweezers experiment were performed at 1064 nm. At this wavelength, most commercially available objectives have around 50% loss [2], which would vastly degrade the squeezed states of light. To improve upon the initial experiments, microscope objectives are required which combine high transmission with high NA. Also, the initial experiments also only provided one-dimensional particle tracking, while many applications require tracking in three dimensions. Future experiments would benefit from combining standard 3D particle tracking techniques [45] with multimode squeezed states [16].

Given the improvements mentioned above, quantum enhanced particle tracking can be expected to perform far better in the near future. Classical particle tracking techniques, however, have limited room for improvement as they are already approaching quantum limited sensitivity [10]. It is reasonable to expect that optical tweezers using squeezed light will soon outperform classical experiments, and lead the field. Considering the rapid rate of technological progress, one may even consider it implausible that in decades to come, particle tracking microscopy should remain confined by the shot-noise limit.

7. Conclusion

Optical tweezers are a powerful tool for biophysical studies which are approaching the limits of classical technology. The quantum technology of squeezed light provides a route to further improved sensitivity. Initial demonstrations have proven that quantum enhanced particle tracking is possible and practical. With further development, quantum enhanced particle tracking is likely to lead the field and enable unprecedented levels of precision in practical biological experiments.

Acknowledgments

This work was supported by the Australian Research Council Discovery Project DP0985078 and the Australian Research Council Centre of Excellence for Engineered Quantum Systems CE110001013.

References

[1] Ashkin A, Dziedzic J M, Bjorkholm J E, and Chu S 1986 Observation of a single-beam gradient force optical trap for dielectric particles *Opt. Lett.* **11** 288–290
[2] Svoboda K and Block S M 1994 Biological applications of optical forces *Annu. Rev. Bioph. Biomol. Struct.* **23** 24785
[3] Bustamante C 2005 Unfolding single RNA molecules: bridging the gap between equilibrium and non-equilibrium statistical thermodynamics *Q. Rev. Biophys.* **38** 291–301
[4] Svoboda K, Schmidt C F, Schnapp B J and Block S M 1993 Direct observation of kinesin stepping by optical trapping interferometry *Nature* **365** 721–7
[5] Finer J T, Simmons R M and Spudich J A 1994 Single myosin molecule mechanics: piconewton forces and nanometre steps *Nature* **368** 113–9
[6] Bustamante C, Chemla Y R, Forde N R and Izhaky D 2004 Mechanical processes in biochemistry *Annu. Rev. Biochem.* **73** 705-48
[7] Greenleaf W J and Block S M 2006 Single-molecule, motion-based DNA sequencing using RNA polymerase *Science* **313** 801
[8] Kukura P, Ewers H, Müller C, Renn A, Helenius A and Sandoghdar V 2009 High-speed nanoscopic tracking of the position and orientation of a single virus *Nat. Methods* **6** 923–9
[9] Moffitt J R, Chemla Y R, Smith S B and Bustamante C 2008 Recent advances in optical tweezers *Annu. Rev. Biochem.* **77** 205-28
[10] Taylor M A, Knittel J and Bowen W P 2013 Fundamental constraints on particle tracking with optical tweezers *New J. Phys.* **15** 023018
[11] Taylor M A, Janousek J, Daria V, Knittel J, Hage B, Bachor H-A and Bowen W P 2013 Biological measurement beyond the quantum limit *Nature Photon.* **7** 229–33
[12] Giovannetti V, Lloyd S and Maccone L 2004 Quantum-enhanced measurements: beating the standard quantum limit *Science* **306**, 1330–6
[13] Kolobov M I and Fabre C 2000 Quantum limits on optical resolution *Phys. Rev. Lett.* **85** 3789–92

[14] Nagata T, Okamoto R, O'Brien J L, Sasaki K and Takeuchi S 2007 Beating the standard quantum limit with four-entangled photons *Science* **316** 726–9
[15] Demkowicz-Dobrzański R, Kołodyński J and Guţă M 2012 The elusive Heisenberg limit in quantum-enhanced metrology *Nat. Commun.* **3** 1063
[16] Treps N *et al.* 2003 A quantum laser pointer *Science* **301** 940–3
[17] Brida G, Genovese M and Ruo Berchera I 2010 Experimental realization of sub-shot-noise quantum imaging *Nature Photon.* **4** 227–30
[18] Abadie J *et al.* 2011 A gravitational wave observatory operating beyond the quantum shot-noise limit *Nature Phys.* **7** 962–5
[19] Crespi A *et al.* 2012 Measuring protein concentration with entangled photons *Appl. Phys. Lett.* **100** 233704
[20] Peterman E J, Gittes F and Schmidt C F 2003 Laser-induced heating in optical traps *Biophys. J.* **84** 1308–16
[21] Neuman K C, Chadd E H, Liou G F, Bergman K and Block S M 1999 Characterization of photodamage to Escherichia coli in optical traps *Biophys. J.* **77** 2856-63
[22] Liu Y, Cheng D K, Sonek G J, Berns M W, Chapman C F and Tromberg B J 1995 Evidence for localized cell heating induced by infrared optical tweezers *Biophys. J.* **68** 2137–44
[23] Lubart R, Lavi R, Friedmann H, and Rochkind S 2006 Photochemistry and photobiology of light absorption by living cells *Photomed. Laser Surg.* **24** 179–85
[24] Lubart R, Wollman Y, Friedmann H, Rochkind S and Laulicht I 1992 Effects of visible and near-infrared lasers on cell cultures *J. Photochem. Photobiol.* **12** 305–10
[25] Da Silva N S and Potrich J W 2010 Effect of GaAlAs laser irradiation on enzyme activity *Photomed. Laser Surg.* **28** 431–4
[26] Zhang F *et al.* 2007Multimodal fast optical interrogation of neural circuitry *Nature* **446** 633–9
[27] Peterman E J, van Dijk M A, Kapitein L C and Schmidt C F 2003 Extending the bandwidth of optical-tweezers interferometry *Rev. Sci. Instrum.* **74** 3246–9
[28] Gittes F, Schnurr B, Olmsted P D, MacKintosh F C and Schmidt C F 1997 Microscopic viscoelasticity: shear moduli of soft materials determined from thermal fluctuations *Phys. Rev. Lett.* **79** 3286–9
[29] Mason T G, Ganesan K, Van Zanten J H, Wirtz D and Kuo S C 1997 Particle tracking microrheology of complex fluids *Phys. Rev. Lett.* **79** 3282–5
[30] Perkins G S and Jones R B 1992 Hydrodynamic interaction of a spherical particle with a planar boundary: II. Hard wall *Phys. A* **189** 447–77
[31] Dufresne E R, Squires T M, Brenner M P and Grier D G 2000 Hydrodynamic coupling of two Brownian spheres to a planar surface *Phys. Rev. Lett.* **85** 3317–20
[32] Meiners J C and Quake S R 1999 Direct measurement of hydrodynamic cross correlations between two particles in an external potential *Phys. Rev. Lett.* **82** 2211–4
[33] Franosch T, Grimm M, Belushkin M, Mor F M, Foffi G, Forró L and Jeney S 2011 Resonances arising from hydrodynamic memory in Brownian motion *Nature* **478** 85–8
[34] Jannasch A, Mahamdeh M and Schäffer E 2012 Inertial Effects of a Small Brownian Particle Cause a Colored Power Spectral Density of Thermal Noise *Phys. Rev. Lett.* **107** 228301
[35] Lukić B, Jeney S, Tischer C, Kulik A J, Forró L and Florin E-L 2005 Direct observation of nondiffusive motion of a Brownian particle *Phys. Rev. Lett.* **95** 160601
[36] Harada Y and Asakura T 1996 Radiation forces on a dielectric sphere in the Rayleigh scattering regime *Opt. Commun.* **124** 529
[37] Gittes F and Schmidt C F 1998 Interference model for back-focal-plane displacement detection in optical tweezers *Opt. Lett.* **23** 7–9
[38] Berg-Sørensen K and Flyvbjerg H 2004 Power spectrum analysis for optical tweezers *Rev. Sci. Instrum.* **75** 594–612
[39] Huang R, Chavez I, Taute K M, Lukić B, Jeney S, Raizen M G and Florin E-L 2011 Direct observation of the full transition from ballistic to diffusive Brownian motion in a liquid *Nature Phys.* **7** 576-80
[40] Li T, Kheifets S, Medellin D and Raizen M G 2010 Measurement of the instantaneous velocity of a Brownian particle *Science* **328** 1673–5
[41] Taylor M A, Knittel J, and Bowen W P 2013 Optical lock-in particle tracking in optical tweezers *Opt. Expr.* **21** 8018
[42] Taylor M A, Janousek J, Daria V, Knittel J, Hage B, Bachor H-A and Bowen W P 2013 Sub-diffraction limited quantum imaging within a living cell *ArXiv*:1305.1353
[43] Stefszky M S *et al.* 2012 Balanced homodyne detection of optical quantum states at audio-band frequencies and below *Class. Quantum Grav.* **29** 145015
[44] Tolić-Nørrelykke I M, Munteanu E-L, Thon G, Oddershede L and Berg-Sørensen K 2004 Anomalous diffusion in living yeast cells *Phys. Rev. Lett.* **93** 078102
[45] Pralle A, Prummer M, Florin E-L, Stelzer E H K, and Hörber J K 1999 Three-dimensional high-resolution particle tracking for optical tweezers by forward scattered light *Microsc. Res. Tech.* **44** 378–386

Quantum Networks with Atoms and Photons

C. Monroe[1], W. Campbell[2], C. Cao[1], T. Choi[1], S. Clark[1], S. Debnath[1], C. Figgatt[1], D. Hayes, D. Hucul[1], V. Inlek[1], R. Islam[1], S. Korenblit[1], K. Johnson[1], A. Manning[1], J. Mizrahi[1], B. Neyenhuis[1], A. Lee[1], P. Richerme[1], C. Senko[1], J. Smith[1], and K. Wright[1]

[1] Joint Quantum Institute, University of Maryland Department of Physics and National Institute of Standards and Technology, College Park, MD 20742

[2] Department of Physics and Astronomy, University of California Los Angeles, Los Angeles, CA 90095

E-mail: monroe@umd.edu

Abstract. Trapped atomic ions are standards for quantum information processing, as all of the fundamental quantum operations have been demonstrated in small collections of atoms. Current work is concentrated on scaling ion traps to larger numbers of interacting qubits and the generation of massive entangled states. We discuss progress in the quantum networking of trapped atomic ions, using the Coulomb interaction for demonstrations of simple quantum simulations of magnetism, ultrafast laser pulses for entanglement, and finally probabilistic photonic interactions to bridge entanglement over long distances.

Trapped atomic ions form ideal quantum systems for the fabrication of quantum information devices and the study of manybody quantum states. Trapped ion qubits enjoy an extreme level of isolation from the environment, they can be entangled through their local Coulomb interaction, and they can be measured with near-perfect efficiency with the availability of cyclic optical transitions [1]. Recent effort in this area has concentrated on the scaling to large numbers of trapped ion qubits, and there is continued progress on the use of monolithic chip traps and integrated optical interfaces for future ion-based quantum devices [2].

Here we highlight new directions in the quantum networking of trapped atomic ion systems undertaken by the JQI-Maryland Ion Trap Group. These efforts do not simply refine and improve earlier approaches of entanglement generation, but represent new methods for the generation of entangled networks with larger systems, at ultrafast times scales, and over macroscopic distances.

In all experiments, we confine atomic $^{171}\mathrm{Yb}^+$ ions in linear radiofrequency (Paul) ion traps, and store effecive spin$-1/2$ qubits in the $^2S_{1/2}$ ground state "clock" hyperfine levels, labeled by $|\uparrow\rangle$ $(F = 1, m_F = 0)$ and $|\downarrow\rangle$ $(F = 0, m_F = 0)$ and separated by a frequency of $\nu_0 = 12.64281$ GHz [3]. The qubits are initialized and detected with laser radiation near resonant with the $^2S_{1/2} - ^2P_{1/2}$ transition at a wavelength around 369.5 nm. We coherently couple the qubit levels via optical stimulated Raman transitions using off-resonant laser radiation. This interaction can be accompanied by optical dipole forces in certain applications.

Ultrafast Quantum Gates

The Raman laser for coherent operations on $^{171}\mathrm{Yb}^+$ qubits is a mode-locked Nd:YVO$_4$ oscillator and amplifier that is frequency tripled to 355 nm. This wavelength is ideal for the $^{171}\mathrm{Yb}^+$ system, as its large detunings from the ^2P states (33 THz blue of $^2P_{1/2}$ and 67 THz red of $^2P_{3/2}$) amount to negligible spontaneous emission and differential AC Stark shifts [4]. Furthermore, the large bandwidth of the pulses easily covers the hyperfine splitting, and the resulting frequency comb from a pulse train allows high resolution resolved-sideband control of trapped ion motion [5], similar to conventional setups involving cw lasers [6].

For higher Raman laser power, we have shown that a solitary pulse with a typical duration of about 10 ps can drive transitions between the two qubit states [4]. Such an interaction ushers a new regime of ultrafast atomic qubit control, and when applied to many trapped ions it has great promise for the operation of ultrafast entangling gates where the gate speed is faster than the period of harmonic trap motion [7, 8].

Toward this end, we apply counterpropagating pulses to the ions, with orthogonal linear polarizations. This creates a spatially modulated Rabi frequency that is the basis of a spin-dependent momentum kick [4]. We have verified the coherence of this process and resulting ultrafast qubit-motion entanglement by implementing a Ramsey interferometer with a single trapped ion. We first create a superposition of spin states with microwaves (first Ramsey $\pi/2$ pulse), then apply two ultrafast spin-dependent kicks with invidual ultrafast laser pulses. After closing the interferometer (second Ramsey $\pi/2$ pulse), we expect fringes only when the kicks are sepated by integral multiples of the trap period, because otherwise the momentum kicks are not parallel, and the motional wavepackets do not recombine. This periodic revival of qubit contrast is shown in Fig 1, showing the coherence of the kick process and the entanglement of spin and motion [9].

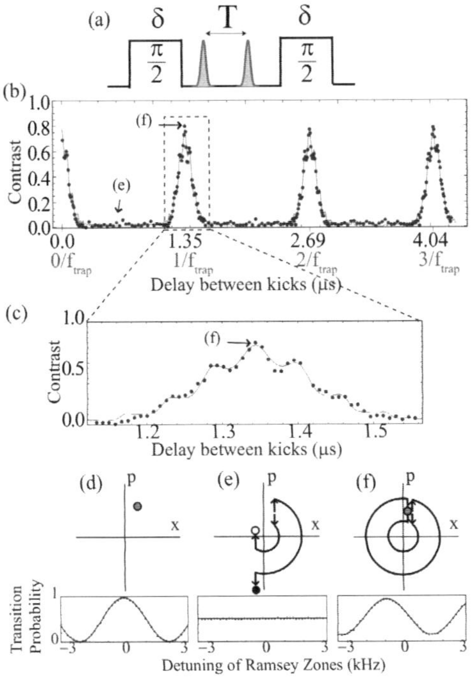

Figure 1. (a) Ramsey sequence for observation of ultrafast spin-dependent momentum kicks. A single trapped ion is driven by two microwave pulses detuned from resonance by δ, with two spin-dependent kicks in between. For each delay T between the kicks, δ is scanned and a Ramsey fringe contrast obtained. (b) Plot of the results of the experiment in (a).s At integer multiples of the trap period, contrast revives. (c) Close-up of the first revival peak in (b). The peak shape is a function of the thermal ion motion, and also residual micromotion that can be seen in the modulation of the revival peak. The best fit curve shown is a fit to theory with free parameters being the micromotion amplitude, average thermal phonon number ($\bar{n} \sim 10$) and maximum contrast revival (80%). (d)-(f): Phase space plots and experimental frequency scans at various delays. The circle color represents spin state: black = $|\downarrow\rangle$, white = $|\uparrow\rangle$. (e) and (f) each correspond to single points in (b), as indicated. (d) No momentum kicks; microwave pulses only. (e) Two kicks separated by half a trap period. (f) Two kicks separated by full trap period. (Figure from Ref. [9].)

Quantum Simulations of Magnetism

Trapped atomic ions can be entangled through the application of spin-dependent dipole forces [10, 11]. Conventionally, such forces are applied to subsets of ions in order to execute entangling quantum gates that are applicable to quantum information processing [1]. When such forces are applied globally, the resulting interacting spin network allows the quantum simulation of a wide variety of spin models such as the Ising and Heisenberg spin chains [12]. The spin-spin interactions originate from modulations of the Coulomb interaction and are therefore long range, and aspects of the system can become classically intractable for even modest numbers of spins $N > 30$. Quantum simulations of magnetism with trapped ions are of particular interest because the interaction graph can be tailored by controlling the external force on the ions, for instance by tuning the spectrum of lasers that provide the dipole force. This allows the control of the sign of the interaction (ferromagnetic vs. antiferromagnetic), the range, the dimensionality, and the introduction of frustration in the system.

Our experiments begin with the ions polarized along an effective magnetic field transverse to long range Ising couplings between the spins [13, 14]. The field is provided by global Raman carrier transitions on all ions, and the Ising couplings arise from an effective Hamiltonian resulting from a virtual coupling through the collective modes of harmonic motion [11]. We operate with the transverse modes of motion in the linear chain [15], so that the modes are tightly packed and all contribute to the effective Ising model. In principle, by controlling the spectrum of light that falls upon each ion in the linear chain, we can program arbitrary fully-connected Ising networks in any dimension [16]. In our current experiments, we use global lasers and a dipole force at a single frequency. However, even with this minimal control we are still able to continuously vary the effective range of the Ising interaction [17]. In practice we can program an Ising interaction that falls off with a power law in distance $1/r^{\alpha}$, with the exponent between $\alpha = 0.5$ and $\alpha = 1.5$. (Theoretically it should be possible to vary the exponent between $\alpha = 0 - 3$ [18, 17].)

Once in an eigenstate (e.g., the ground state) of the transverse field, we experimentally ramp the field down, and given its final value, we then turn off the interactions and directly measure the state of each spin with a CCD camera. If the field was ramped down adiabatically, we expect the resulting spin order to reflect the properties of the more interesting Ising couplings of the Hamiltonian. For ferromagnetic couplings, we observe a clear phase transition from polarization along the transverse field, and recognize a sharpening in the transition point near the critical magnetic field, as the number of spins is increased [19]. For antiferromagnetic couplings, the gaps are much smaller and the adiabaticity criterion is more stringent. Fig. 2 shows the order seen following a ramp to zero field for $N = 10$ ions, where excited states are clearly created. Of the $2^{10} = 1024$ possible configurations, the two degenerate staggered-order ground states are produced about 18% of the time, with low-lying excitations also prevalent, as shown in the figure. We also observe that as the interaction is made longer range, the creation of excited states is more likely, as the increased level of frustration closes the gap even further and hinders adaiabatic evolution. The magnetic field ramp is controlled with a simple radiofrequency modulation source, and we can easily program any type of evolution in time. In addition to using standard exponential and linear ramps, we have also calculated the optimum ramp shape that maintains a constant adiabaticity parameter throughout the evolution, provided through prior knowledge of the many body energy spectrum [20].

By adding an axial field to an antiferromagnetic Ising Hamiltonian, we expect there to be many different orders to emerge as the axial field competes with the Ising couplings. We have been able to directly observe the steps between these orders by lowering the transverse field as before, but to a final state where only the Ising couplings and axial field remain. For N spins, we expect to see $[N/2]$ distinct phases in the ground state, owing to the long-range nature of the Ising interaction. Moreover, as N grows to infinity, the "Devil's staircase" structure [21] in magnetization becomes a fractal that arises since every rational filling factor (of which there are infinitely many) is a ground state [22].

We have performed quantum simulations with up to $N = 18$ trapped ion spins. While such quantum simulations can (barely) be predicted classically with under 20 spins, this system holds great promise for the quantum simluation of fully-connected spin chains with 50−100 ions, where many aspects of the system such as dynamical processes become intractable.

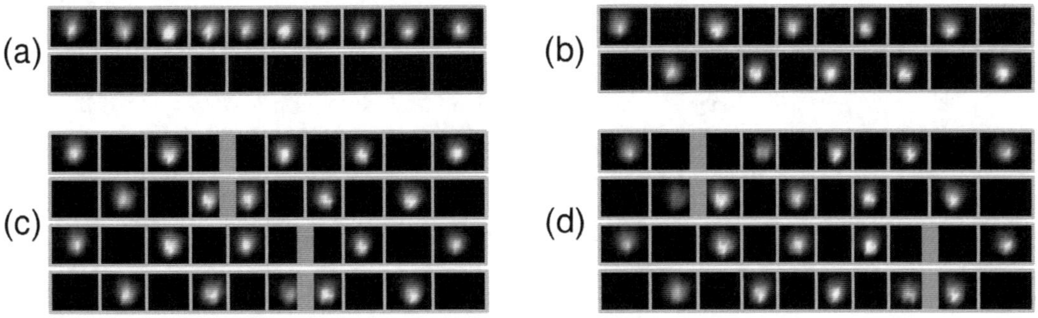

Figure 2. Magnetic ordering of $N = 10$ trapped atomic ion spins with long range antiferromagnetic Ising coupling, conveyed by the spatial images of the crystal having with a length of $22\mu m$. (a) Reference images of all spins prepared in state $|\uparrow\rangle$ (top) and $|\downarrow\rangle$ (bottom). (b) Ground state staggered order of spins after adiabatically lowering transverse magnetic field. These two degenerate states are produced a total of $\sim 18\%$ of the time. (c) Lowest (coupled) excited states, showing domain walls near the center of the crystal (in green). Any of these four degenerate states are produced a total of $\sim 4\%$ of the time. (d) Next lowest (coupled) excited states, showing domain walls near the ends of the crystal (in green), with these four degenerate states produced a total of $\sim 2\%$ of the time.

Photonic Quantum Networks

Apart from the Coulomb interaction, trapped atomic ions can also be entangled over remote distances through photonic interfaces. Here, an emitted photon from a trapped atomic ion can propagate while carrying entanglement with its host atom, through its polarization, frequency, or longitudinal degrees of freedom [23]. When photons emerging from multiple trapped ions are made to interfere on a beamsplitter, subsequent detection of the photons can herald the entangement of the trapped ion qubits. While such an interface is necessarily probabilistic, it is not post-selective, and such an interface can be scaled and be made fault-tolerant [24].

There are two common protocols for entanglement of the host ions, involving either one or two photons emitted from the ion pair. In the single photon protocol, each ion is weakly excited, and the beamsplitter erases which path information on which ion emitted the photon, and when the photon is detected on either detector the ions are projected into an entangled state. The excitation is kept weak so that the probability of having two photons is small enough [25]. However, this introduces a fundamental source of error in the entanged state, and moreover this scheme requires interferometric stability of the path lengths to better than an optical wavelength, including the recoiling ions. In the two-photon protocol, the ions are both excited with near unit probability, and the two emitted photons interfere in such a way that only the singlet entangled state of the photons results in their emerging on opposite ports of the beamsplitter [26]. The two photon protocol does not suffer from the shortcomings of the single photon protocols, and for good enough collection probabilities, it will always outperform the single photon protocol. Both protocols can be enhanced by the use of optical cavities to increase the probability of light collection, but this does not change the probabilitistic nature of the interface (despite unrealistic proposals to determinsitically use photons to entangle ions).

Our early experiments coupling ions to free space admitted a pairwise entanglement rate of about 0.001 Hz [27], limited by the low photon collection probability and also the slow initialization of the ion before each attempt. Current experiments have significantly enhanced the entanglement rate to approximately 1 Hz, owing to improvements in the particular protocol, initialization, optical light collection, fiber coupling, and detector efficiencies. In the future, the possibility of integrating photonics with ion traps, especially on a semiconductor or silicon chip platform, holds great promise to enhance this rate even further, perhaps by also using optical cavities to further enhance the light collection efficiency [28]. In the end, a juxtaposition of both Coulomb entanglement techniques and photonic entanglement protocols will likely be necessary to scale up the ion trap system [2]. Here, local deterministic operations can be performed within small ion crystals, accompanied by distant-independent probabilistic operations for scaling to very large quantum systems.

Acknowledgments

This work is supported by the U.S. Army Research Office (ARO) Award W911NF0710576 with funds from the DARPA Optical Lattice Emulator Program, ARO award W911NF0410234 with funds from the IARPA MQCO program, ARO award W911NF0910406 with funds from a MURI on Quantum Optical Circuits, and the NSF Physics Frontier Center at JQI.

References

[1] Wineland D and Blatt R 2008 *Nature* **453** 1008–1014
[2] Monroe C and Kim J 2013 *Science* **339** 1164
[3] Olmschenk S, Younge K C, Moehring D L, Matsukevich D N, Maunz P and Monroe C 2007 *Phys. Rev. A* **76** 052314
[4] Campbell W C, Mizrahi J, Quraishi Q, Senko C, Hayes D, Hucul D, Matsukevich D N, Maunz P and Monroe C 2010 *Phys. Rev. Lett.* **105** 090502
[5] Hayes D, Matsukevich D N, Maunz P, Hucul D, Quraishi Q, Olmschenk S, Campbell W, Mizrahi J, Senko C and Monroe C 2010 *Phys. Rev. Lett.* **104** 140501
[6] Liebfried D, Blatt R, Monroe C and Wineland D 2003 *Reviews of Modern Physics* **75** 281
[7] García-Ripoll, J J, Zoller, P and Cirac J I 2003 *Phys. Rev. Lett.* **91** 157901
[8] Duan L M 2004 *Phys. Rev. Lett.* **93** 100502
[9] Mizrahi J, Senko C, Neyenhuis B, Johnson K G, Campbell W C, Conover C W S and Monroe C 2013 *Phys. Rev. Lett.* **110** 203001
[10] Cirac J I and Zoller P 1995 *Phys. Rev. Lett.* **74** 4091–4094
[11] Mølmer K and Sørensen A 1999 *Phys. Rev. Lett.* **82** 1835
[12] Porras D and Cirac J I 2004 *Phys. Rev. Lett.* **92** 207901
[13] Friedenauer A, Schmitz H, Glueckert J T, Porras D and Schaetz T 2008 *Nature Physics* **4** 757–761
[14] Kim K, Chang M S, Islam R, Korenblit S, Duan L M and Monroe C 2009 *Phys. Rev. Lett.* **103** 120502
[15] Zhu S L, Monroe C and Duan L M 2006 *Phys. Rev. Lett.* **97** 050505
[16] Korenblit S, Kafri D, Campbell W C, Islam R, Edwards E E, Gong Z X, Lin G D, Duan L M, Kim J, Kim K and Monroe C 2012 *New Journal of Physics* **14** 095024
[17] Islam R, Senko C, Campbell W C, S K, Smith J, Lee A, Edwards E E, Wang C C J, Freericks J K and Monroe C 2013 *Science* **340** 583
[18] Britton J, Sawyer B, Keith A, Wang C C J, Freericks J, Uys H, Beircuk M and Bollinger J 2012 *Nature* **484** 489–492
[19] Islam R, Edwards E, Kim K, Korenblit S, Noh C, Carmichael H, Lin G D, Duan L M, Wang C C J, Freericks J and Monroe C 2011 *Nature Communications* **2:377**
[20] Richerme P, Senko C, Smith J, Lee A, Korenblit S and Monroe C (*Preprint* arXiv 1305.2253)
[21] Bak P and Bruinsma R 1982 *Phys. Rev. Lett.* **49** 249–251
[22] Richerme P, Senko C, Korenblit S, Smith J, Lee A, Islam K R, Campbell W C and Monroe C (*Preprint* arXiv 1303.6983)
[23] Luo L, Hayes D, Manning T A, Matsukevich D N, Maunz P, Olmschenk S, Sterk J D and Monroe C 2009 *Fortschr. Phys.* **57** 1133
[24] Duan L M and Monroe C 2010 *Reviews of Modern Physics* **82** 1209
[25] Cabrillo C, Cirac J I, García-Fernández P and Zoller P 1999 *Phys. Rev. A* **59** 1025–1033
[26] Simon C and Irvine W T M 2003 *Phys. Rev. Lett.* **91** 110405
[27] Moehring D L, Maunz P, Olmschenk S, Younge K C, Matsukevich D N, Duan L M and Monroe C 2007 *Nature* **449** 68
[28] Sterk J D, Luo L, Manning T A, Maunz P and Monroe C 2012 *Phys. Rev. A* **85** 062308

Journal of Physics: Conference Series **467** (2013) 012009 doi:10.1088/1742-6596/467/1/012009

An ultra-high optical depth cold atomic ensemble for quantum memories

B. M. Sparkes[1], J. Bernu[1], M. Hosseini[1], J. Geng[1,2], Q. Glorieux[1,3], P. A. Altin[4], P. K. Lam[1], N. P. Robins[4], and B. C. Buchler[1].

[1] Centre for Quantum Computation and Communication Technology, Department of Quantum Science, Research School of Physics and Engineering, The Australian National University, Canberra, ACT 0200, Australia

[2] Quantum Institute for Light and Atoms, Department of Physics, East China Normal University, Shanghai 200062, P. R. China

[3] Joint Quantum Institute, University of Maryland, College Park, Maryland 20742, USA

[4] Quantum Sensors Lab, Department of Quantum Science, Australian National University, Canberra, ACT 0200, Australia

E-mail: `ping.lam@anu.edu.au`

Abstract. Quantum memories for light lie at the heart of long-distance provably-secure communication. Demand for a functioning quantum memory, with high efficiency and coherence times approaching a millisecond, is therefore at a premium. Here we report on work towards this goal, with the development of a ^{87}Rb magneto-optical trap with a peak optical depth of 1000 for the D2 $F = 2 \rightarrow F' = 3$ transition using spatial and temporal dark spots. With this purpose-built cold atomic ensemble we implemented the gradient echo memory (GEM) scheme on the D1 line. Our data shows a memory efficiency of $80 \pm 2\%$ and coherence times up to 195 μs.

1. Introduction

Quantum repeaters are needed to extend the distance of quantum key distribution past its current limit of hundreds of kilometres. A key component of a quantum repeater is a quantum memory. A quantum memory for this application needs to have an efficiency approaching unity without adding noise to the state, and storage times from milliseconds to seconds. Ideally it would also have a high bandwidth and be able to store many pieces of information simultaneously. Many different techniques are currently being developed and much experimental progress has been made over the past few years, with: efficiencies approaching 90% [1]; storage times of over seconds [2, 3]; bandwidths of gigahertz [4, 5]; and over 1000 pieces of information stored at one time [6]. These results were, however, achieved using different memory schemes in different storage media. The challenge now is to reproduce these results with one memory.

Here we will focus on the gradient echo memory (GEM) scheme, which shows great promise due to the high efficiencies achieved in both warm atomic vapours [1] and solid state systems [7]. GEM has also been used to demonstrate temporal compression and stretching of pulses, as well as a capacity to arbitrarily resequence stored information [8] and the interference of initially time-separated pulses [9, 10]. However, the maximum coherence time for GEM to date is 50 μs [11], limited by atomic diffusion in warm vapors.

Motivated by other work demonstrating both high efficiencies [12] and long storage times [3] in cold atomic systems, we introduce a cold atom realisation of GEM. To achieve this we first had to develop an ultra-high optical depth (OD) atomic source to allow for high efficiency recall.

2. Developing the Cold Atomic Source

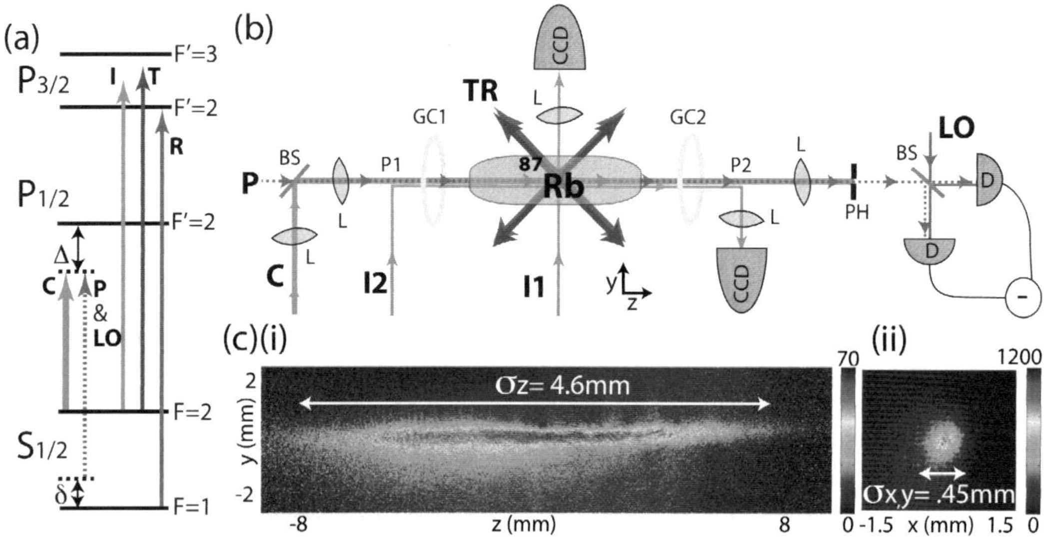

Figure 1. (a) Energy level diagram for all fields present: C - coupling; P - probe; LO - local oscillator; I - Imaging; T - trapping; R - repump. Also shown are one- and two-photon detunings for probe and coupling fields (Δ and δ respectively). (b) Experimental set-up, with: ^{87}Rb - atomic ensemble; BS - 50:50 beam-splitter; L - lens; D - photo-diode detector; I1(2) - imaging beam 1(2); CCD - charged-coupled device camera; PH - pin-hole; GC - GEM coils; P1 and P2 - positions for inserting mirrors for axial (z) imaging with I2. MOT coil configuration not shown, neither is vertical (y) MOT beam. (c)(i) Side-on and (ii) front-on images of the optimised atomic ensemble. Colour bars show OD scales, σ values show the standard deviation of Gaussian fits to ensemble.

The work-horse for cold atom experiments is the magneto-optical trap (MOT), and we used a rubidium-87 MOT as our storage medium. For quantum memory applications we aimed to produce a MOT with low atomic temperatures and a very large OD. To create as high an OD as possible we optimised both the static loading of the MOT through geometry, a spatial dark spot and optical de-pumping, followed by a compression phase using a temporal dark spot. After describing the application of these techniques, we present a characterisation of the system using near-resonance absorption imaging.

We used a basic three-beam retro-reflection configuration, with the frequencies of all lasers involved illustrated in Fig. 1(a). The trapping and cooling laser had a total power of 400 mW after spatial filtering and was red-detuned by 30 MHz from the D2 $F = 2 \rightarrow F' = 3$ transition for the loading of the MOT. The repumping field was on resonance with the D2 $F = 1 \rightarrow F' = 2$ transition.

The optimal shape for a cold atomic ensemble for use as a quantum memory is a cylinder along the direction of the memory beams to allow for maximum absorption of the probe. To create this shape while still allowing easy access for the memory beams we used four elongated coils in a quadrupole configuration to create a 2D MOT in the z direction (memory axis) and positioned the horizontal MOT beams at 45° to the long axis of the MOT. An extra set of axial coils in the z direction created 3D confinement. The shape of the MOT could then be determined by the currents in the 2D and axial coils, as well as the relative intensities of the trapping fields.

For the loading phase, the cylindrically symmetric magnetic field gradient produced by the 2D MOT coils was 16 G/cm, and for the axial coils was 2 G/cm. The ^{87}Rb atoms were produced from a natural-mixture Rb dispenser inside a $100\times50\times50$ mm^3 single-sided, anti-reflection coated cell. This cell was attached to a vacuum system consisting of a 70 L/s ion pump supplemented by a passive titanium sublimation pump; with the dispenser running in the cell we measure a background pressure at the ion pump of 1.5×10^{-9} kPa.

The density in the trapped atomic state ($F = 2$) was limited by reabsorption of fluorescence photons within the MOT (leading to an effective outwards radiation pressure [13]). By placing a dark spot of approximately 7.5 mm in diameter in the repump, atoms at the centre of the MOT were quickly pumped into the lower ground state ($F = 1$) and became immune to this unwanted effect, allowing for a higher density of atoms in the centre of the trap, as first demonstrated in [14].

We could collect over 10^{10} atoms in this configuration. However, we found that the static MOT parameters that optimise atom number did not optimise density in the compression and cooling phase of the MOT, and we typically worked with a sample of 4×10^9 atoms.

In this second stage of the ensemble preparation we used a temporal dark spot to transiently increase the density of the sample by simultaneously increasing the trapping laser and repump detunings and increasing the magnetic field gradient [15, 16, 17]. Detuning the trapping laser created some of the conditions for polarisation-gradient cooling (PGC) [18], which could be used to achieve much colder and denser ensembles than in a standard MOT. We smoothly ramped the frequency of the trapping beam from 30 to 70 MHz below resonance, and the repump beam to 8 MHz below resonance, over a period of 20 ms. The 2D magnetic field gradient in the x and y directions was ramped up to 40 G/cm as the trapping and repump lasers are detuned. We did not ramp the axial field.

Finally, we optically pumped the atoms into the desired ground state ($F = 1$, see Fig. 1(a)) by simply turning off the repump 50 μs before the trapping beam.

Absorption imaging was used to optimise and characterise the MOT. The set-up and frequency of the imaging beams are shown in Fig. 1. We performed imaging both across (I1) and along (I2) the z axis (in which case two mirrors are temporarily placed at locations P1 and P2).

As the absorption of light by atoms away from resonance will follow a Lorentzian decay, to be able to have a precise value for OD it was important to have a well calibrated line centre. This is especially important as one goes further off resonance as the relation between on-resonance OD and off-resonance OD depends on the one-photon detuning (Δ) as follows:

$$\mathrm{OD}_{res} = \frac{\Delta^2 + \gamma^2/4}{\gamma^2/4} \ln(I_t/I_o), \tag{1}$$

where I_t is the transmitted imaging beam intensity passing through the MOT off resonance, I_o is the intensity measured without the MOT present, γ is the excited state decay rate. For Rb γ is approximately 6 MHz. To measure Δ accurately we lowered the atom number in our trap until the OD did not saturate on resonance and plotted out the resonance curve as a function of detuning to accurately locate the line centre.

An image of the optimised MOT is shown in Fig. 1(c). For this image we used a 4.98 s load time followed by 20 ms of ramping fields. All fields were then turned off and an image of the

MOT was taken 500 μs later, with a comparison image being taken 150 ms later to obtain as precise a measure of I_o as possible while ensuring no atoms were still present. As the imaging beam was on the closed D2 $F = 2 \rightarrow F' = 3$ transition, it was necessary to pump atoms back into the $F = 2$ state before these images were taken. This was achieved with a 200 μs pulse of on-resonance repump light immediately before the image. The front-on image was taken 60 MHz detuned, and the side-on image was taken 20 MHz detuned from resonance. The temperature of the MOT was approximately 200 μK.

3. Gradient Echo Memory Using Cold Atoms

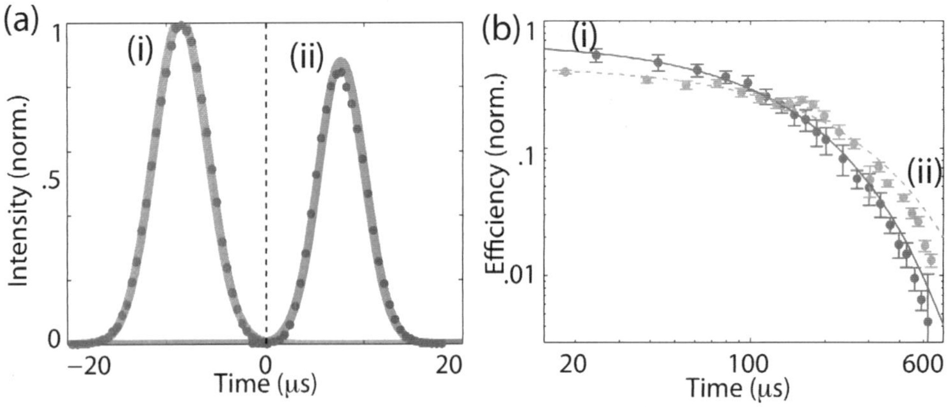

Figure 2. (a) High efficiency demodulated and squared heterodyne data: (i) input pulse (blue), and (ii) echo (red) at $80 \pm 2\%$ total efficiency. Points correspond to digitally demodulated data, averaged over 10 (17) traces for input (echo) and squared, error from standard deviation. Lines correspond to Gaussian fit. Dashed line indicates magnetic field switching at $t = 0$. (b) Memory efficiency vs storage time for storage (i) 500 μs (blue, solid) and (ii) 3 ms after MOT fields off (green, dashed).

The details of the GEM scheme are covered in depth in previous papers [1, 8, 19, 20, 21]. Here we used the three-level implementation [19], with a weak probe field and a strong coupling field, and switching magnetic field gradients produced via two solenoids, one placed either side of the MOT. This set-up and the frequencies of the fields are shown in Fig. 1(b). In the three-level implementation the effective OD is reduced by the detuning of the laser, and by the Zeeman broadening of the transition, with the final effective OD being nearly 1000-fold reduced from the on-resonance OD. This is why a high OD ensemble is critical for high efficiencies in GEM.

For the highest efficiency memory we used a 480 ms load time, and implemented the memory protocol 500 μs after the MOT fields had been turned off. We found a Gaussian pulse with a full-width-half-maximum of 10 μs to be optimal for storage. For this pulse length, a one-photon detuning of -250 MHz and approximately 350 μW in the coupling field (corresponding to a Rabi frequency of 2 MHz), we were able to demonstrate storage with $80 \pm 2\%$ total efficiency. This was measured using heterodyne detection and is shown in Fig. 2(a). As heterodyne detection is mode sensitive, care was taken to optimise the visibility for the input pulse, so that any change in the mode during storage would lead to a reduction in the measured efficiency. The coherence time for this set-up was determined to be 117 μs by storing the pulses for longer periods of time (Fig. 2(b)). By implementing the memory protocol 3 ms after the MOT had been turned off, this was increased to 195 μs.

Further improvements to the efficiency of GEM using a cold atomic ensemble could be achieved by increasing the MOT OD further, for instance by optical pumping all atoms into the correct m_F ground state. Currently the main decoherence mechanisms are not fully understood, but potentially could include inhomogeneous background magnetic fields generated via eddy currents, as well as atomic diffusion. Improving the switching electronics and physically redesigning the coil configuration could help to improve the former, while cooling the ensemble further could improve the latter.

For more information on GEM using cold atoms, see Ref. [22].

4. Conclusion

Here we have presented the development of a cold atomic ensemble with a peak OD of 1000 on the $D2$ $F = 2 \rightarrow F' = 3$ transition, and demonstrated a memory efficiency of $80 \pm 2\%$ using GEM, the highest to date using cold atoms. The primary limit on this efficiency is still the OD, which could be improved by optically pumping all the atoms into the correct m_F state. The decoherence of the system was found to be exponential, with a time constant of 117-195 μs. This is a factor of 2-4 greater than in warm vapour GEM experiments. To improve upon this, further investigations would be required to determine the limiting factors. Even without further improvement, however, our memory provides an excellent high-OD platform for numerous other proof-of-principle experiments.

Acknowledgments

Thanks to Colin Dedman for his help designing the fast switching GEM coils. This research was conducted by the *Australian Research Council Centre of Excellence for Quantum Computation and Communication Technology* (project number CE110001027). NPR is supported by an Australian Research Council QEII Fellowship. QG is supported by the AFSOR and the Physics Frontier Center at the NIST/UMD Joint Quantum Institute.

References

[1] Hosseini M, Sparkes B M, Campbell G, Lam P K and Buchler B C 2011 *Nature Commun.* **2** 174
[2] Longdell J J, Fraval E, and Sellars M J, and Manson N B 2005 *Phys. Rev. Lett.* **95** 063601
[3] Dudin Y O, Li L and Kuzmich A 2013 Light storage on the time scale of a minute *Phys. Rev. A* **87**, 031801
[4] Saglamyurek E *et al.* 2012 *Phys. Rev. Lett.* **108**, 083602
[5] Reim K F *et al.* 2010 *Nature Photon.* **4** 218-221
[6] Bonarota M, Le Gouët J-L and Chanelière T 2011 *N. J. Phys.* **13** 013013
[7] Hedges M P, Longdel J J, Li Y and Sellars M J 2010 *Nature* **465** 1052-1056
[8] Hosseini M, Sparkes B M, Hétet G, Longdell J J, Lam P K and Buchler B C 2009 *Nature* **461** 241-245
[9] Campbell G, Hosseini M, Sparkes B M, Lam P K and Buchler B C 2012 *N. J. Phys* **14** 033022
[10] Sparkes B M, Hosseini M, Cairns C, Higginbottom D, Campbell G T, Lam P K and Buchler B C 2012 *Phys. Rev. X* **2** 021011
[11] Higginbottom D, Sparkes, B M, Rancic M, Pinel O, Hosseini M, Lam P K and Buchler B C 2012 *Phys. Rev. A* **86** 023801
[12] Chen Y-H, Lee M-J, Wang I, Du S, Chen Y-F, Chen Y-C and Yu, I A 2012 *Phys. Rev. Lett.* **110** 083601
[13] Metcalf H J and van der Straten P 2003 *J. Opt. Soc. Am. B* **20** 887-908
[14] Ketterle W, Davis K B, Joffe M A, Martin A and Pritchard, D E 1993 *Phys. Rev. Lett.* **70** 2253-2256
[15] Lee H J, Adams C S, Kasevich M and Chu S 1996 *Phys. Rev. Lett.* **76** 2658-2661
[16] DePue M T, Winoto S L, Han D J and Weiss D S 2000 *Opt. Commun.* **180** 73-79
[17] Petrich W, Anderson M H, Ensher J R, Cornell E A 1994 *J. Op. Soc. Am. B* **11** 1332-1335
[18] Dalibard J and Cohen-Tannoudji C 1989 *J. Opt. Soc. Am. B* **6** 2023-2045
[19] Hétet G, Hosseini M, Sparkes B M, Oblak D, Lam P K and Buchler B C 2008 *Opt. Lett.* **33** 2323-2325
[20] Hosseini M, Campbell G, Sparkes B M, Lam P K and Buchler B C 2011 *Nature Phys.* **7** 794-798
[21] Hosseini M, Sparkes B M, Campbell G T, Lam P K, and Buchler B C 2012 *J. Phys. B* **45** 124004
[22] Sparkes B M, Bernu J, Hosseini M, Geng J, Glorieux Q, Altin P A, Lam P K, Robins N P and Buchler B C 2013 *N. J. Phys* **15**, 085027

Probing local quantities in a strongly interacting Fermi gas

Yoav Sagi, Tara E. Drake, Rabin Paudel, Roman Chapurin and Deborah S. Jin

JILA, National Institute of Standards and Technology and the University of Colorado, and the Department of Physics, University of Colorado, Boulder, CO 80309-0440, USA

E-mail: `jin@jilau1.colorado.edu`

Abstract. The collective behavior of an ensemble of strongly interacting fermions is central to many physical systems, and its theoretical description is challenging due to the many-body nature of the problem. The ultracold Fermi gas is an ideal model system to shed light on this issue, as it provides excellent controllability, reproducibility, and unique detection methods. One of the problems, however, which complicates the interpretation of such experiments is the inherent density inhomogeneity of the gas due to harmonic confinement. We have developed a technique to overcome this difficulty by selectively probing atoms near the center of a trapped gas while still retaining momentum resolution. In this contribution to the 21th International Conference on Laser Spectroscopy (ICOLS 2013), we give an overview of this technique and some of the observations that have resulted from its implementation.

1. Introduction

When a gas of fermions is cooled to a temperature (T) lower than the Fermi temperature (T_F), Fermi-Dirac statistics start to have an appreciable effect [1]. At $T = 0$, a homogeneous gas of non-interacting fermions fills all available states in momentum space up to the Fermi surface. The sharpness of the transition from filled states to empty states (the Fermi step) is determined by T/T_F. When the density of the gas is non-uniform in space but the variations are gradual enough (i.e. the local density approximation holds), then the different densities at each position give rise to different local T_F. This is the case in all experiments where the atoms are trapped by some non-uniform external potential; in most cases, this potential is approximately harmonic. Since the cloud is in thermal equilibrium (T is the same everywhere), T/T_F will vary across the gas. Any physical quantity of interest which depends on the scaled temperature, $Q(T/T_F)$, will also vary across the gas. A global measurement of that quantity will yield

$$\overline{Q} = N^{-1} \int Q\left[T/T_F(\mathbf{r})\right] n(\mathbf{r}) d^3r, \tag{1}$$

where $n(\mathbf{r})$ is the density distribution, and N is the total number of atoms. In general, it can be difficult to interpret the measured $\overline{Q}$ and compare it directly to theoretical predictions. Moreover, this density-weighted averaging causes sharp features in $Q(T/T_F)$ to wash-out and disappear in $\overline{Q}$.

A prime example of such a sharp feature is the step in the momentum distribution of non-interacting fermions at low T/T_F. In a harmonically trapped gas, the highest density, and therefore also the lowest T/T_F, occurs in the center. The outskirts of the cloud are effectively at higher temperature (higher T/T_F), which means that the corresponding Fermi step is broader. When averaged together, the resulting momentum distribution of a trapped cloud does not exhibit a sharp transition and the location of the transition shifts. This situation is illustrated in figure 1.

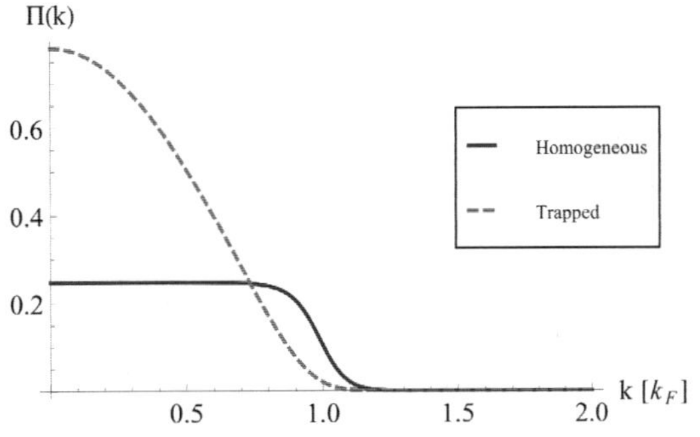

Figure 1. The wash-out of sharp features due to density inhomogeneity of a trapped gas. The momentum distribution, $\Pi(k)$, of a homogeneous non-interacting Fermi gas (solid blue line) compared to that of a harmonically trapped gas (red dashed line), both at $T/T_F = 0.1$. In both cases, we normalize the distribution such that $\int \Pi(\mathbf{k})d^3k = 1$. The momentum is given in units of the Fermi momentum, k_F; either the trap k_F for the trapped gas, given by $(2m/\hbar)^{1/2}(\omega_x\omega_y\omega_z 6N)^{1/6}$, or the homogeneous k_F, given by $(6\pi^2 n)^{1/3}$. Here, m is the mass, ω_i is the oscillation frequency in the trap for axis i, n and N are the density and total number of atoms in one of the two interacting spin states, respectively.

For a strongly interacting Fermi gas, much attention has been devoted in the past to studies of the Fermi superfluid and the BCS-BEC crossover. At unitarity ($1/k_F a \to 0$, where a is the scattering length), the transition temperature is around $0.2T_F$ [2]. Above this temperature, the gas is in the normal phase, the nature of which is still under debate. Some theories predict that pairing instability will persist into the normal phase at temperatures close to the phase transition, leading to a pseudogap phase [3, 4, 5, 6]. Experiments on a trapped gas using atoms photoemission spectroscopy (PES) have found evidence for the existence of this phase [2]. The range of temperatures at which a pseudogap phase is expected to exist is very limited [5], hence signatures of its existence are easily obscured by averaging over inhomogeneous density, which includes parts of the cloud with much higher T/T_F. As such, there is a need to obtain quantitative results on a strongly interacting Fermi gas with a temperature spread as small as possible.

The experiments described here were performed with an ultra-cold gas of ^{40}K atoms [7, 8, 9]. The atoms are initially collected from a high temperature isotopically enriched source by a magneto-optical trap, after which they are pushed by a laser to a second chamber with a much lower residual pressure, where they are collected in a second magneto-optical trap. Evaporative cooling is performed initially in a magnetic trap, followed by evaporation in a crossed dipole trap. The evaporation ends with an equal mixture of the $|f, m_f\rangle = |9/2, -9/2\rangle$ and $|9/2, -7/2\rangle$ spin states, where f is the quantum number denoting the total atomic spin and m_f is its projection. The scattering length between these two states can be controlled by an applied magnetic field via the s-wave Fano-Feshbach resonance around 202 G. The final stage of the evaporation is done in a magnetic field of 203.4 G where the s-wave scattering length between the two spin states is about $-850a_0$, where a_0 is the Bohr radius. Typically, there are around 100000-250000 atoms per spin state. We detect the atoms using absorption imaging after a time of flight ranging between 3 ms to 16 ms.

2. The crossed hollow light beams technique

Ideally, a solution to the density inhomogeneity problem should do the following: reduce the spread of T/T_F in the sample being probed, enable a measurement of the momentum distribution of the probed sample (e.g. PES can be performed), and not limit the range of achievable temperatures and densities. The latter requirement makes solutions based on modifying the trapping geometry challenging, since they may result in lower densities which makes the evaporation less efficient. However, much improvement has been reported recently in this route, where experiments have demonstrated the creation of a Bose-Einstein condensate in a quasi-uniform potential [10]. We have chosen not to alter the trapping geometry, but to post-select a small fraction of the atoms near the cloud center for interrogation [8]. In the local density approximation, a small volume of the gas behaves as a homogeneous system.

Probing atoms in a small volume is accomplished by intersecting two orthogonally propagating hollow light beams that optically pump atoms into a spin state that is dark to the detection. The spatial mode of the optical pumping beams is a second-order Laguerre-Gaussian, given by $I(r) = \frac{P}{\pi w^2}\left(\frac{2r^2}{w^2}\right)^2 e^{-\frac{2r^2}{w^2}}$, where P is the total optical power and w is the waist. The fraction of atoms probed can be changed by varying the intensity of these optical pumping beams. Usually, the two beams are aligned to the center of the cloud (i.e. where T/T_F is lowest and the density is highest), but in principle, it is possible to probe any position in the cloud. It is important to keep in mind that the atoms which are not probed are not driven out of the trap, but only optically pumped to a dark state. The hollow light beams are turned on close to the time of the trap release, where initially the beam that propagates perpendicular to the long axis of the cloud is turned on for $10\,\mu$s followed by a $40\,\mu$s pulse of the second beam. These durations are much shorter than all relevant dynamical timescales in the experiment.

3. Experiments with a homogeneous Fermi gas

The full, step-like momentum distribution of a homogeneous gas of non-interacting fermions is shown in most statistical mechanics textbooks but has never been observed directly. The momentum distribution of a trapped gas of non-interacting fermions was measured before, and its shape is very well understood, albeit not displaying any sharp features [11]. In a first set of experiments, we used the hollow light beams technique to directly observe the Fermi surface [8]. In these experiments, the final magnetic field is set to 208.2 G, where the scattering length is approximately $-30\ a_0$ (a_0 is the Bohr radius) and the gas is very weakly interacting with a dimensionless interaction strength of $1/k_F a \sim 91$. The hollow light beams are switched on immediately after the trap is switched off. Absorption images are taken after 10 ms or 12 ms of time-of-flight. The three-dimensional momentum distribution is obtained from the integrated two-dimensional image by applying an inverse Abel transform.

As can be seen in figure 2, as the intensity of the hollow light beams is increased, a momentum distribution with a sharp step gradually emerge. When the fraction of atoms probed is less than 30%, we find that the data fits very well to a homogeneous Fermi-Dirac distribution [8]. Moreover, from that fit we can extract the average k_F and temperature of the gas, which correspond to the position and sharpness of the step. We find that sharpness of the Fermi step in our data at low probing fractions is limited by the temperature and not by the remaining density inhomogeneity.

In the regime of strong interactions, we have used the hollow light beam technique to measure Tan's contact [9]. The contact has emerged in recent years as an important thermodynamic parameter for quantum gases that connects many seemingly unrelated quantities through a set

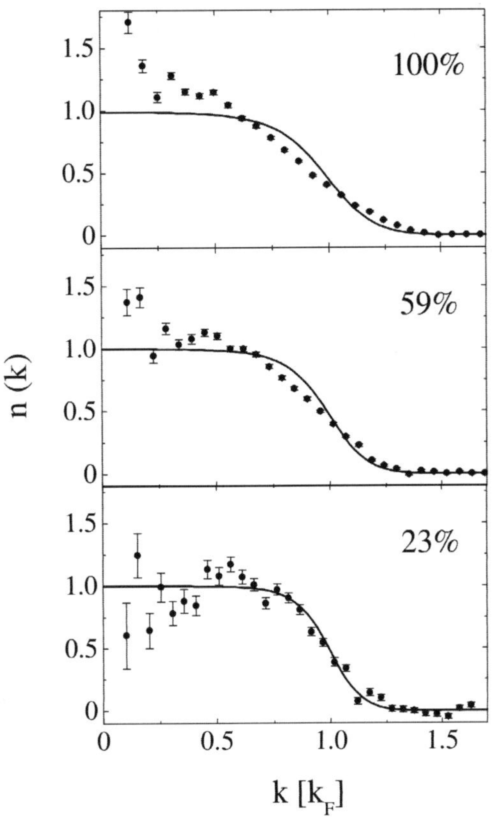

Figure 2. The emergence of the Fermi surface in a gas of non-interacting fermions. The normalized momentum distributions for the central 100%, 59%, and 23% of the trapped gas. The solid lines show fits to the expected momentum distribution for a homogeneous Fermi gas; we fix T to the value determined for the trapped gas, leaving k_F (and the corresponding T_F) as single fit parameter. The momentum is given in units of this k_F. This data was taken at approximately $T/T_{F,trap} = 0.12$.

of universal relations [12, 13, 14, 15, 16, 17]. Many of these relations were tested experimentally [18, 19, 20, 21]. The dependence of the contact on temperature was investigated in several theoretical works [22, 23, 24, 25, 26]. Interestingly, there are significant discrepancies in the predicted contact of a homogeneous system from different theories, especially in the vicinity of the superfluid phase transition. This makes the homogeneous contact a good benchmark for many-bodies theories. For a trapped gas, these discrepancies almost completely vanish because of averaging over different local T/T_F [27].

We extract the contact from measurement of the rf line shape, which measures the probability that an rf pulse at some frequency transfers an atom from one of the two interacting spin states to a third weakly interacting state. A universal relation connects the high frequency tail of the line shape to the contact, C (see Ref. [17] and references therein): $\Gamma(\nu)/\int_{-\infty}^{\infty}\Gamma(\nu')d\nu' = C/\left(Nk_F\sqrt{2}\pi^2\nu^{3/2}\right)$ for $\nu \to \infty$, where ν is the rf detuning in units of E_F/h, (E_F is the Fermi energy and h is Planck's constant). After cooling the gas, the magnetic field is ramped in 50 ms from 203.4 G to the Feshbach resonance at 202.2 G. The hollow light beams are turned on 280 μs before trap release. The line shape is measured using a 100 μs long rf pulse with a gaussian field envelope with a width of 17 μs. This pulse is centered 180 μs before trap release. We set the rf power well below the value where we see the onset of saturation of the number of outcoupled atoms, such that the line shape is indeed measured in the linear response regime.

As with a trapped gas, the tail of the measured line shape of a homogeneous sample shows the expected scaling of $\nu^{-3/2}$, from which we extract the homogeneous contact [9]. We have found a gradual decrease of the contact when the temperature is increased, in good agreement with the virial expansion (which is expected to hold for $T/T_F \gtrsim 1$). We have also observed a sharp drop in the contact at a temperature consistent with the superfluid phase transition. Several theoretical works predict a narrow enhancement in the contact around the superfluid critical temperature, due to pair fluctuations [23]. However, we do not observe a such pronounced peak in the experiment.

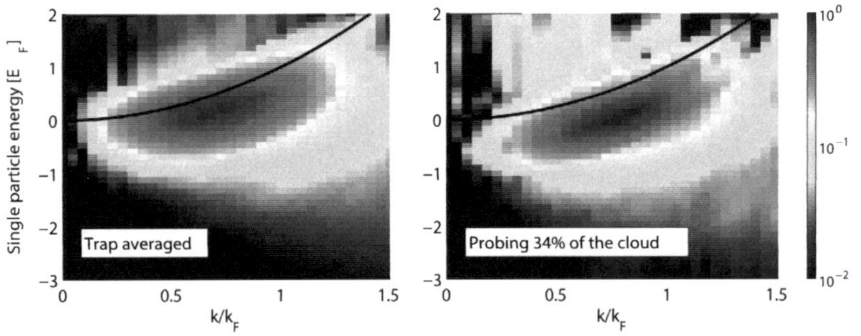

Figure 3. PES measurement on a homogeneous unitary Fermi gas. The occupied spectral function, with and without applying the hollow light beams technique, of a unitary gas ($1/k_F a = 0$) at $T/T_c = 0.8$, where T_c is the superfluid phase transition, determined using the molecular projection technique [28, 2]. The color at each energy and momentum is proportional to the probability to excite from the many-body state an atom with that particular energy and momentum (plotted in a logarithmic scale). As a comparison, the black solid line is the quadratic dispersion of a non-interacting particle.

A measurement technique that can probe the many-body state of the gas and provide information about the nature of the normal state is PES [7]. Atom PES measurement is achieved using momentum-resolved spectroscopy. Since the outcoupled atoms are weakly interacting, their energy-momentum dispersion is simply that of a free particles. Using this fact and energy and momentum conservation, it is possible to reconstruct the occupied spectral function of system, which gives the probability to find an atom at some energy and momentum [7]. As in the case of the measurement of the Fermi surface, at trap release we pulse the hollow light beams such that the spectral function is measured only for a small fraction of the gas at the cloud center.

An example of such a measurement is shown in figure 3, for a gas at $T/T_c = 0.8$, where T_c is the superfluid critical temperature. The left panel shows the measured occupied spectral function of a trapped gas, while the right panel shows the result when we probe only the central 34% of the gas. Several observations can be made; much of the signal that has been removed from the left panel is close to the free-particle line. This is expected, since the atoms in the outskirts of the cloud are at much lower densities (and higher effective T/T_F), so the mean-field interaction energy shift is much smaller. Also, the widths of the energy distribution curves (EDC), which are vertical cuts through the data, are narrower for the more homogeneous sample on the right. We expect future homogeneous PES measurements at different interaction strengths and temperatures in the BCS-BEC crossover to provide insight on the nature of the normal state.

4. Conclusion and outlook

The combination of of both spatial and momentum resolution in the hollow light beams technique enable us to probe important quantities in a homogeneous sample. There is a tradeoff between how homogeneous the sample we probe is and the signal-to-noise ratio in our measurement, and we find that probing $\sim 30\%$ of the cloud seems to strike the right balance. We have presented here several measurements using the hollow light beams technique, both for weakly and strongly interacting fermions. Future projects include a thorough study of the homogeneous atom PES across the BCS-BEC crossover, and investigation of the spatial variance of the condensate fraction below the superfluid phase transition.

We would like to thank the JILA BEC group for fruitful discussions. We acknowledge funding from the NSF and NIST.

References

[1] DeMarco, B., Papp, S. B., and Jin, D. S. *Phys. Rev. Lett.* **86**, 5409–5412 Jun (2001).
[2] Gaebler, J. P., Stewart, J. T., Drake, T. E., Jin, D. S., Perali, A., Pieri, P., and Strinati, G. C. *Nat. Phys.* **6**(8), 569–573 August (2010).
[3] Chien, C.-C., Guo, H., He, Y., and Levin, K. *Phys. Rev. A* **81**, 023622 Feb (2010).
[4] Hu, H., Liu, X.-J., Drummond, P. D., and Dong, H. *Phys. Rev. Lett.* **104**, 240407 Jun (2010).
[5] Magierski, P., Wlazłowski, G., and Bulgac, A. *Phys. Rev. Lett.* **107**, 145304 Sep (2011).
[6] Perali, A., Palestini, F., Pieri, P., Strinati, G. C., Stewart, J. T., Gaebler, J. P., Drake, T. E., and Jin, D. S. *Phys. Rev. Lett.* **106**, 060402 Feb (2011).
[7] Stewart, J. T., Gaebler, J. P., and Jin, D. S. *Nature* **454**(7205), 744–747 August (2008).
[8] Drake, T. E., Sagi, Y., Paudel, R., Stewart, J. T., Gaebler, J. P., and Jin, D. S. *Phys. Rev. A* **86**, 031601 Sep (2012).
[9] Sagi, Y., Drake, T. E., Paudel, R., and Jin, D. S. *Phys. Rev. Lett.* **109**, 220402 Nov (2012).
[10] Gaunt, A. L., Schmidutz, T. F., Gotlibovych, I., Smith, R. P., and Hadzibabic, Z. *Phys. Rev. Lett.* **110**, 200406 May (2013).
[11] Jin, D. S. and Regal, C. A. *Ultra-cold Fermi Gases*, volume 164 of *Proceedings of the International School of Physics "Enrico Fermi"*, chapter Fermi Gas Experiments, 1–51. IOS Press (2007).
[12] Tan, S. *Ann. Phys.* **323**, 2952–2970 (2008).
[13] Tan, S. *Ann. Phys.* **323**, 2987–2990 (2008).
[14] Tan, S. *Ann. Phys.* **323**, 2971–2986 (2008).
[15] Braaten, E., Kang, D., and Platter, L. *Phys. Rev. Lett.* **100**, 205301 (2008).
[16] Zhang, S. and Leggett, A. J. *Phys. Rev. A* **79**, 023601 (2009).
[17] Braaten, E. In *The BCS-BEC Crossover and the Unitary Fermi Gas*, Zwerger, W., editor, volume 836 of *Lecture Notes in Physics*, 193–231. Springer Berlin / Heidelberg (2012).
[18] Stewart, J. T., Gaebler, J. P., Drake, T. E., and Jin, D. S. *Phys. Rev. Lett.* **104**, 235301 Jun (2010).
[19] Kuhnle, E. D., Hu, H., Liu, X.-J., Dyke, P., Mark, M., Drummond, P. D., Hannaford, P., and Vale, C. J. *Phys. Rev. Lett.* **105**, 070402 Aug (2010).
[20] Partridge, G. B., Strecker, K. E., Kamar, R. I., Jack, M. W., and Hulet, R. G. *Phys. Rev. Lett.* **95**, 020404 Jul (2005).
[21] Werner, F., Tarruell, L., and Castin, Y. *Eur. Phys. J. B* **68**, 401–415 (2009).
[22] Yu, Z., Bruun, G. M., and Baym, G. *Phys. Rev. A* **80**, 023615 Aug (2009).
[23] Palestini, F., Perali, A., Pieri, P., and Strinati, G. C. *Phys. Rev. A* **82**, 021605 Aug (2010).
[24] Enss, T., Haussmann, R., and Zwerger, W. *Ann. Phys.* **326**(3), 770 – 796 (2011).
[25] Hu, H., Liu, X.-J., and Drummond, P. D. *New J. Phys.* **13**(3), 035007 (2011).
[26] Drut, J. E., Lähde, T. A., and Ten, T. *Phys. Rev. Lett.* **106**, 205302 May (2011).
[27] Kuhnle, E. D., Hoinka, S., Dyke, P., Hu, H., Hannaford, P., and Vale, C. J. *Phys. Rev. Lett.* **106**, 170402 Apr (2011).
[28] Regal, C. A., Greiner, M., and Jin, D. S. *Phys. Rev. Lett.* **92**, 040403 Jan (2004).

AUTHOR INDEX

Institute of Physics
Dirac House, Temple Back
Bristol BS1 6BE UK

CURRAN ASSOCIATES INC
proceedings
.com